AF734867

Additive Manufacturing of Thin-Walled Structures

This comprehensive book examines the manufacturing of thin-walled structures using diverse Additive Manufacturing (AM) processes. It emphasizes fundamental principles alongside cutting-edge developments in AM, covering materials ranging from non-metallic to metallic thin-wall applications. This book provides complete coverage from pre-processing through post-processing stages, explores both traditional and machine learning-based characterization techniques for thin-walled structures, and presents real-world case studies demonstrating various engineering applications of thin-walled designs.

Key features

- Exclusively focuses on AM of thin-walled structures, including fabrication, characterization, and simulation
- Highlights quality control and post-processing of thin-walled structures
- Features simulation models used to study the complex distribution of thermal stresses in thin-walled structures
- Explains parametric modeling and process parameter optimization for AM of quality thin-walled structures
- Sheds light on the applications of additive-manufactured thin-walled structures

This book is aimed at graduate students and researchers in manufacturing engineering, AM, and modeling and simulation.

Advanced Materials Processing and Manufacturing

Series Editor:

Kapil Gupta

The CRC Press Series in Advanced Materials Processing and Manufacturing covers the complete spectrum of materials and manufacturing technology, including fundamental principles, theoretical background, and advancements. Considering the accelerated importance of advances for producing quality products for a wide range of applications, the titles in this series reflect the state-of-the-art in understanding and engineering the materials processing and manufacturing operations. Technological advancements for enhancement of product quality, process productivity, and sustainability, are on special focus including processing for all materials and novel processes. This series aims to foster knowledge enrichment on conventional and modern machining processes. Micro-manufacturing technologies such as micro-machining, micro-forming, and micro-joining, and Hybrid manufacturing, additive manufacturing, near net shape manufacturing, and ultra-precision finishing techniques are also covered.

Advanced Welding Technologies
Bridging Theory, Practice and Innovations in Fusion, Solid-State and Beyond
Edited by Bappa Acherjee, Dawei Zhao and Fábio A.O. Fernandes

Process Modeling and Optimization in Modern Manufacturing
Edited by Ravi Pratap Singh, Narendra Kumar, Vishal Francis and Ankur Jaiswal

Additive Manufacturing for Biomedical Implants and Prosthetics
Edited by Pawan Kumar Rakesh and Pravin Kumar

Manufacturing in the Digital Age
Edited by Mohan Kumar Pradhan, Mahendra U. Gaikwad, and Ragavanantham Shanmugam

Fused Deposition Modelling
Technology, Applications, and Developments
Edited by Anshuman Kumar Sahu and Siba Sankar Mahapatra

Manufacturing Tribology
Edited by Jibin T Philip, Chacko Preno Koshy, and MD Mathew

Machine Learning for Advanced Manufacturing
Edited by Nishant Ranjan, Rashi Tyagi, Ranvijay Kumar, Ashutosh Tripathi, and Amit Verma

Engineered Ceramics and Composite Materials
From Fundamentals to Applications
Edited by Neethu Bhaskar, Sathish Kumar Palaniappan, Sanjay Mavinkere Rangappa, and Suchart Siengchin

Additive Manufacturing of Thin-Walled Structures
Edited by Sagar Nikam, Mayur Sawant, and Shaun McFadden

For more information about this series, please visit: www.routledge.com/Advanced-Materials-Processing-and-Manufacturing/book-series/CRCAMPM

Additive Manufacturing of Thin-Walled Structures

Edited by Sagar Nikam, Mayur Sawant,
and Shaun McFadden

CRC Press
Taylor & Francis Group
Boca Raton London New York

CRC Press is an imprint of the
Taylor & Francis Group, an **informa** business

Designed cover image: Shutterstock

First edition published 2026
by CRC Press
2385 NW Executive Center Drive, Suite 320, Boca Raton FL 33431

and by CRC Press
4 Park Square, Milton Park, Abingdon, Oxon, OX14 4RN

CRC Press is an imprint of Taylor & Francis Group, LLC

© 2026 selection and editorial matter, Sagar Nikam, Mayur Sawant and Shaun McFadden; individual chapters, the contributors

Reasonable efforts have been made to publish reliable data and information, but the author and publisher cannot assume responsibility for the validity of all materials or the consequences of their use. The authors and publishers have attempted to trace the copyright holders of all material reproduced in this publication and apologize to copyright holders if permission to publish in this form has not been obtained. If any copyright material has not been acknowledged please write and let us know so we may rectify in any future reprint.

Except as permitted under U.S. Copyright Law, no part of this book may be reprinted, reproduced, transmitted, or utilized in any form by any electronic, mechanical, or other means, now known or hereafter invented, including photocopying, microfilming, and recording, or in any information storage or retrieval system, without written permission from the publishers.

For permission to photocopy or use material electronically from this work, access www.copyright.com or contact the Copyright Clearance Center, Inc. (CCC), 222 Rosewood Drive, Danvers, MA 01923, 978-750-8400. For works that are not available on CCC please contact mpkbookspermissions@tandf.co.uk

For Product Safety Concerns and Information please contact our EU representative GPSR@taylorandfrancis.com. Taylor & Francis Verlag GmbH, Kaufingerstraße 24, 80331 München, Germany.

Trademark notice: Product or corporate names may be trademarks or registered trademarks and are used only for identification and explanation without intent to infringe.

ISBN: 9781032915302 (hbk)
ISBN: 9781032945248 (pbk)
ISBN: 9781003571209 (ebk)

DOI: 10.1201/9781003571209

Typeset in Times
by codeMantra

Contents

About the Editors

Sagar Nikam is working as a Lecturer at Ulster University (Derry~Londonderry Campus). He completed his PhD from the Indian Institute of Technology Indore. Thereafter, he joined the Birla Institute of Technology Mesra as an Assistant Professor. Before joining as a Lecturer in Mechanical Engineering in the School of Computing, Engineering, and Intelligent Systems at Ulster University, Magee Campus, he worked as a Postdoctoral Research Associate on developing simulations for the selective laser melting process. His research interest is in additive manufacturing with a particular interest in developing process simulation models for metal additive manufacturing processes. He is also a Fellow of the Higher Education Academy (FHEA) and Chartered Engineer (CEng) registered with IMechE: Institution of Mechanical Engineers. He has experience in publishing quality research articles in journals, conferences, and book chapters. All of these can be accessed through the link https://scholar.google.co.in/citations?user=MitnY6EAAAAJ&hl=en.

Mayur Sawant completed his PhD from the Indian Institute of Technology Indore. Thereafter, he joined the Indian Institute of Technology Indore as a postdoctoral fellow. He was awarded an MHRD-sponsored Shastri Indo-Canadian Postdoctoral Fellowship to pursue research at the University of Alberta, Canada. He later joined as an Assistant Professor in the Department of Manufacturing Engineering and Industrial Management at COEP Technological University, Pune. He is currently working as an Assistant Professor in the Department of Mechanical Engineering, School of Engineering and Science, MIT Art, Design and Technology University. He has been working in various research fields such as additive manufacturing, hybrid additive manufacturing, biomaterials, lattice structures, and knee implants. He has experience in publishing quality research articles in journals, conferences, and book chapters. All of these can be accessed through the link https://scholar.google.com/citations?user=bPxgPecAAAAJ&hl=en.

Shaun McFadden completed his PhD from University College Dublin (2007) in the area of computational microstructure prediction during the solidification of alloys, with a particular focus on macrostructure development under microgravity conditions. He held academic positions at the Technological University of Dublin and at Trinity College Dublin (University of Dublin) before joining Ulster University in 2015, where he is a Senior Lecturer in the School of Computing, Engineering, and Intelligent Systems. He has published widely in areas related to materials science, both fundamental and practical, and has experience conducting and supervising research in the areas of metal additive manufacturing, microstructure modeling, crystal growth experimentation, and joining technology for thin-walled structures. He is a Fellow of the Higher Education Academy (FHEA) and a Chartered Engineer (CEng) registered with Engineers Ireland. His publication track record can be accessed via the link https://scholar.google.com/citations?user=zBdDFrQAAAAJ&hl=en.

Contributors

K. G. Anbarasu
Department of Mechanical Engineering, SRM Institute of Science and Technology, Vadapalani, Chennai, Tamil Nadu, India

K. G. Ashok
Department of Mechanical Engineering, Easwari Engineering College Ramapuram, Chennai, Tamil Nadu, India

Tejaswini Bhosale
Department of Computer Science and Engineering, MIT Art, Design and Technology University, Pune, India

Dermot Brabazon
School of Mechanical and Manufacturing Engineering, Dublin City University, Glasnevin Campus, Dublin, Ireland

Abhijit Cholkar
South East Advanced Manufacturing (SEAM) Research Centre, Applied Technology Building, South East Technological University-Waterford, Waterford, Ireland

Shree Choudhari
School of Mechatronics Engineering, Symbiosis Skills and Professional University, Pune, India

Andrea El Hassanin
Department of Chemical, Materials and Industrial Production Engineering, University of Naples Federico II, Piazzale Tecchio, Naples, Italy

N. Harshavardhana
Department of Mechanical Engineering, SRM Institute of Science and Technology, Kattankulathur, Chengalpattu, India

Neelesh Kumar Jain
Department of Mechanical Engineering, Indian Institute of Technology Indore, Simrol, MP, India

Sarika Kumari
Department of Biotechnology, Central University of South Bihar, Gaya (Bihar), India

R. Majusha
Department of Electrical and Electronics Engineering, St peter's College of Engineering and Technology, Avadi, Chennai, Tamil Nadu, India

M. Manjaiah
Department of Mechanical Engineering, National Institute of Technology, Warangal, Telangana, India

Manoj Mugale
Eastern Kentucky University, Richmond, Kentucky, USA

Shubhavardhan Ramadurga Narasimharaju
Department of Mechanical and Manufacturing, South East Technological University, Cork road main campus, Waterford, Ireland

Akriti Nigam
Department of Computer Science and Engineering, Birla Institute of Technology Mesra, Ranchi, India

Deepika Nikam
School of Computing, Engineering, and Intelligent Systems, Ulster University, Magee Campus, Northland Road, Londonderry, Northern Ireland, UK

Sagar Nikam
School of Computing, Engineering, and Intelligent Systems, Ulster University, Magee Campus, Northland Road, Derry/Londonderry, Northern Ireland, UK

Pralhad Pesode
Department of Mechanical Engineering, Dr. Vishwanath Karad MIT-World Peace University, Pune, India

Ramesh Raghavendra
South East Advanced Manufacturing (SEAM) Research Centre, Applied Technology Building, South East Technological University-Waterford, Waterford, Ireland

M. Satthiyaraju
Department of Mechanical Engineering, SRM Institute of Science and Technology Vadapalani, Chennai, Tamil Nadu, India

Mayur Sawant
Department of Mechanical Engineering, MIT Art, Design and Technology University, Pune, India

Fabio Scherillo
Department of Chemical, Materials and Industrial Production Engineering, University of Naples Federico II, Piazzale Tecchio, Naples, Italy

S. P. Sundar Singh Sivam
Department of Mechanical Engineering, SRM Institute of Science and Technology, Kattankulathur, Chengalpattu, India

Antonino Squillace
Department of Chemical, Materials and Industrial Production Engineering, University of Naples Federico II, Piazzale Tecchio, Naples, Italy

Antonio Viscusi
Department of Engineering and Sciences, Faculty of Technology and Innovation Sciences, Universitas Mercatorum, Piazza Mattei, Rome, Italy
Faculty Member of Department of Chemical, Materials and Industrial Production Engineering, University of Naples Federico II, Piazzale Tecchio, Naples, Italy

Sagar Wankhede
School of Mechatronics Engineering, Symbiosis Skills and Professional University, Pune, India

Preface

Additive manufacturing (AM) is redefining modern engineering by enabling the fabrication of components with complex geometries and tailored material properties. Among its many transformative applications, the production of **thin-walled structures** stands out for its ability to deliver components that are lightweight, efficient, and structurally optimized for use in aerospace, automotive, biomedical, and architectural sectors. However, despite this promise, the AM of thin-walled structures presents numerous technical challenges.

This book provides a structured exploration of the latest advancements, challenges, and innovations in the AM of thin-walled structures, following a logical progression through nine dedicated chapters.

Chapter 1 introduces the fundamental role of additively manufactured thin-walled structures, examining their importance, applications, and the unique advantages they offer over traditionally manufactured counterparts. This sets the stage for **Chapter 2**, which addresses the key challenges associated with AM of thin-walled components such as thermal residual stresses, metallurgical defects, anisotropic properties, and stability. With this context established, **Chapter 3** explores the specific AM processes used to fabricate thin-walled structures across metals, polymers, and ceramics. It offers a comparative analysis of techniques such as selective laser melting, direct metal laser sintering, electron beam melting, and wire arc additive manufacturing, with an emphasis on process optimization and material-specific considerations. Building on process understanding, **Chapter 4** presents advanced modeling and simulation approaches. This chapter examines thermal models, empirical methods, and finite element simulations that enable the prediction of melt pool behavior, deposition characteristics, and residual stresses, providing crucial insights for process refinement and component design. Then progressing to **Chapter 5**, which focuses on data-driven techniques for feature measurement and defect monitoring. The integration of machine learning and deep learning is explored as a powerful toolset for real-time defect detection, process control, and the implementation of intelligent manufacturing systems aligned with Industry 4.0. In **Chapter 6**, attention shifts to the characterization of thin-walled structures. Given their sensitivity to geometric imperfections and nonlinear behavior, accurate characterization is important. This chapter examines state-of-the-art experimental techniques and evolving standards (ASTM, ISO) that support reliable assessment of mechanical and structural performance. Recognizing that as-built AM components typically require additional refinement, **Chapter 7** addresses the post-processing techniques essential for improving surface quality, dimensional accuracy, and mechanical integrity. This chapter covers primary and secondary post-processing methods, including heat treatments and surface finishing. This discussion continues in **Chapter 8**, which delves into advanced post-processing techniques such as hot isostatic pressing, laser polishing, chemical treatments, mechanical polishing, and hybrid methods. These treatments are critical for achieving the high-performance standards demanded by industrial applications. Finally, **Chapter 9** explores the rapidly evolving field of bimetallic thin-walled structures. By combining the properties of different metals, these structures offer unparalleled opportunities for performance optimization. This chapter also addresses sustainability concerns, the role of high-entropy alloys and functionally graded materials, and the integration of AI-driven process enhancements.

Through this cohesive flow, this book seeks to provide both a deep technical foundation and a forward-looking perspective on the AM of thin-walled structures. It is intended to serve as a resource for researchers, engineers, and students who are shaping the next generation of lightweight, high-performance components.

1 Introduction to Additively Manufactured Thin-Walled Structures

Pralhad Pesode, Manoj Mugale, and Sagar Wankhede

1.1 INTRODUCTION

Additive manufacturing (AM), widely known as 3D printing [1–3], has revolutionized the fabrication of three-dimensional (3D) components by enabling the creation of intricate geometries directly from computer-aided design (CAD) models through a layer-by-layer approach [4–6]. This innovative technique has proven particularly advantageous for industries such as aerospace, biomedical engineering, shipbuilding, construction, and automotive [7–9], where optimizing material usage and achieving lightweight structures without compromising mechanical properties are critical. Weight reduction and material efficiency have been central objectives in these fields for decades, driving extensive research into lightweight materials and advanced fabrication techniques [10]. Traditional solutions, such as composites and aluminum alloys, have provided incremental improvements [11]. However, their ability to push the boundaries of material performance remains constrained. For instance, the density of the lightest conventionally fabricated materials typically exceeds 8–10 g/cm^3, as further reductions often result in a significant compromise in mechanical properties [12,13].

AM addresses the challenges of producing complex structures by enabling the freeform fabrication of designs that were previously unattainable using conventional methods such as stamping, forging, casting, rolling, and drawing. This includes thin-walled geometries and intricate lattice structures, which are key to enhancing the strength-to-weight ratio of components. Thin-walled geometries are characterized by walls with thicknesses significantly lower than the length and height of the component. The development of such geometries, along with lightweight metal components, is a prominent research area in AM. Lattice structures, a subset of these geometries, are particularly challenging to fabricate using conventional processes. Thin-walled and hollow-shaped structures are particularly susceptible to stability loss over time. Stability constraints are critical considerations in applications such as fuselage and rocket structures. Consequently, it is essential to evaluate the mechanical properties, material characteristics, design

DOI: 10.1201/9781003571209-1

factors, challenges, and limitations when fabricating thin-walled structures using AM techniques. Similarly, parts manufactured for biomedical applications, such as scaffolds and engineered tissues, serve as excellent examples of this category [14–17]. However, AM has made it possible to directly construct these spatial structures with controlled and low porosity, making it a primary target sector for the technology. With its ability to facilitate rapid prototyping and customization, AM has been widely adopted for producing lightweight and highly functional components, including topologically optimized designs and intricate metallic lattice structures [18].

The ability of AM to precisely fabricate thin-walled objects with thicknesses as small as 85 μm is one of its most impressive accomplishments [19]. With the help of this capacity, engineers can achieve the strict industry requirements for sustainability and performance while optimizing material utilization and mechanical integrity. While many applications involving AM fabrication of thin-walled structures rely on powder-based fusion techniques, wire feedstock-based methods offer several notable advantages. Metallic wires are easier to manufacture compared to metal powders and are widely available. AM techniques that utilize energy sources such as electric arcs, lasers, plasmas, and electron beams are well suited for processing wire feedstock. Among these, laser and electron beam-based methods stand out for their ability to achieve high precision in fabrication. However, the resolution of the existing AM methods limits the design of thin-walled structures. While wire-fed machines have resolutions of more than 500 μm, powder bed systems usually offer resolutions of about 100 μm [20]. Even though powder bed systems offer finer resolution, the spatial precision is often unpredictable, and due to the intrinsic constraints of the AM process, items produced within this size range typically exhibit lower mechanical strength. The impact of decreasing wall thickness on the mechanical characteristics of manufactured specimens, specifically, Ti-6Al-4V made by electron beam melting (EBM) and SS-304L made by selective laser melting (SLM), was examined by Ben et al. [21]. Their research determined the mechanical property thresholds for these materials in powder bed AM systems that mark the change from bulk specimens to thin-walled structures.

Moreover, advancements in AM have facilitated the exploration of strategies inspired by nature, emulating the complex architectures of biological materials like bones and shells. These natural structures demonstrate an exceptional balance of lightness, strength, and stiffness through hierarchical designs. By replicating these features, AM enables the creation of materials with customizable and superior mechanical properties, far surpassing the capabilities of conventional materials. As a result, AM not only enhances the lightweight potential of structures but also redefines manufacturing paradigms, allowing for unparalleled design freedom and application-specific innovations. This technology continues to play a pivotal role in advancing engineering solutions across diverse sectors, pushing the limits of what is achievable in material science and structural optimization.

Metal thin-walled structures are characterized by wall thickness significantly smaller than their other dimensions. While there is no strict numerical definition for "thin-walled," these structures are defined by their mechanical behavior, where bending deflections dominate over transverse shear deformations. This makes them ideal for applications requiring reduced weight and material usage without compromising structural integrity. These structures are widely used in structural applications, encountering various stresses such as compression, bending, shear, torsion, or their combinations. They exhibit unique buckling and post-buckling behaviors under these loads, enabling them to function even in highly stressed environments. This makes them invaluable in aerospace fuselages, automotive frames, and other load-bearing applications. However, thin-walled structures face challenges related to thermal effects during fabrication. Deformation due to varying cooling rates after high-temperature deposition can compromise geometrical stability and dimensional accuracy. Thinner parts are particularly susceptible to thermal accumulation due to shorter scanning paths and lower thermal conductivity. To address these issues, controlling cooling rates and implementing compensation design strategies are crucial for minimizing deformation and maintaining accuracy [22].

Fabrication of polymeric thin-walled and hollow structures is primarily achieved using extrusion and photopolymerization-based AM techniques. Overhanging features often require support structures to prevent sagging, which increases build time, reduces surface quality, and necessitates additional post-processing, driving up costs. Strategies such as multi-directional setups, custom extrusion paths, and varying build directions have been proposed to minimize the need for support materials [23]. Key challenges include real-time control of polymer melt solidification and evaporation during fabrication. Proper energy deposition in light- and laser-based AM techniques is crucial for achieving defect-free geometries. Peng et al. [24] demonstrated the use of a mist spray setup to accelerate solidification and mitigate deformation. Despite these challenges, thin-walled parts fabricated via AM generally exhibit satisfactory mechanical properties.

Ceramic thin-walled structures are primarily fabricated using laser-based deposition and photopolymerization, with the latter focusing on microfabrication and the design of thin-walled structures. After depositing and photocuring ceramic slurry into multilayer tracks, heat treatment ensures stability, mechanical properties, and surface quality. Techniques like lateral overlap compensation improve layer uniformity, but challenges such as surface roughness (ladder effects) limit photopolymerization for small and intricate features [25]. Laser-based AM, a growing technology for ceramics, offers fast and cost-effective fabrication. However, issues like micro-cracks, density variations, and color changes arise due to high processing temperatures. Controlling laser power and optimizing parameters can mitigate these challenges, thereby improving surface quality and mechanical properties to match those of conventionally manufactured parts [14]. Additionally, heat treatments such as sintering can restore the color of the feedstock [26].

Thin-walled structures are an inherently efficient design strategy, as they minimize dead load while maximizing performance. Their high strength-to-weight ratio makes them indispensable in industries such as aerospace and automotive, where lightweight components are critical to improving efficiency, reducing fuel consumption, and enhancing overall performance. A wide variety of geometrical configurations have been explored to harness the potential of thin-walled structures. These include plates, tubes, hollow spheres, and shell structures. Among these, tubular structures have garnered significant attention due to their exceptional energy absorption capabilities, which are achieved through progressive plastic deformation. This makes them ideal for crashworthiness applications in automotive design and impact-resistant components in aerospace engineering. Thin-walled structures also find application in energy and thermal management systems, such as heat exchangers, where their lightweight and high-surface-area configurations enhance heat transfer efficiency. In biomedical engineering, thin-walled scaffolds are used to mimic natural bone structures, providing mechanical support while facilitating tissue regeneration.

Overall, the AM-manufactured thin-walled structure helps lower production costs and resource consumption; the ability to produce components with thinner walls also denotes a significant improvement in the efficiency of engineering systems. As a result, there is a noticeable decrease in fuel consumption, which in turn reduces the volume of hazardous gases released into the atmosphere. The trend of lighter thin-walled components is essentially a significant step toward future transportation and industrial processes that are more environmentally friendly and sustainable [27]. This work presents a specialized and application-driven perspective by emphasizing the critical role of thin-walled structures in industries such as aerospace, automotive, biomedical, and architecture. It systematically addresses the fabrication challenges, mechanical property limitations, and process constraints unique to ultra-thin geometries across metals, polymers, and ceramics. Unifying insights across different AM techniques highlights the impacts of process resolution, thermal deformation, and mechanical instability on component quality. The following section discusses the various types of materials used to produce thin-walled structures by various AM techniques.

1.2 AM METHODS AND MATERIALS FOR THIN-WALLED STRUCTURES

1.2.1 AM Methods

Figure 1.1 presents the classification of AM techniques based on material type and resolution. For metal thin-walled structures, powder bed fusion (PBF) and directed energy deposition (DED) are the most utilized AM processes [19,28]. Within PBF, the two predominant methods are laser powder bed fusion (LPBF) (e.g., SLM, direct metal laser sintering (DMLS)) and EBM. For DED, laser metal deposition (LMD) (e.g., laser engineered net shaping (LENS), directed light fabrication, direct metal deposition, and laser cladding) and wire arc additive manufacturing (WAAM) are the most recognized techniques [29]. Apart from these material jetting [30], binder jetting [6], and material extrusion [31] methods, other techniques are also used for metal processing.

The extrusion-based AM processes (e.g., fused deposition modeling (FDM) or fused filament fabrication) are most commonly used/explored for low-melting-point materials [32,33]. Polymer-based composites can be fabricated via selective laser sintering (SLS), vat photopolymerization, and PBF, among other methods [34]. A few commonly used manufacturing techniques for thin-walled structures are briefly explained as follows.

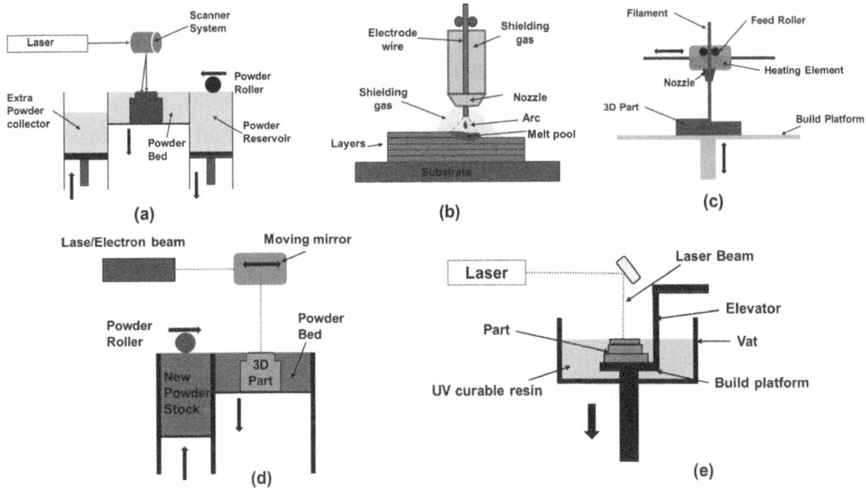

FIGURE 1.1 Various AM methods: (a) SLS, (b) WAAM, (c) FDM, (d) PBF, and (e) SLA. AM, additive manufacturing; SLD, selective laser sintering; WAAM, wire arc additive manufacturing; FDM, fused deposition modeling; SLA, stereolithography.

1.2.1.1 Powder Bed Fusion

PBF is an AM process that utilizes thermal sources, such as laser beams, to fuse powdered particles predictably, enabling the design of objects using CAD. PBF has two predominant methods: LPBF (e.g., SLM, SLS, and DMLS) and EBM [30]. In SLM, a high-power laser is used to provide heat and fully melt and fuse the metallic powder. Usually using inert gas, the operation takes place in a controlled setting to avoid oxidation. Metal powders, including stainless steel, titanium and its alloys, and aluminum alloys, are typically subjected to the procedure. Because SLM can create lightweight, robust, and geometrically complex components directly from CAD drawings, it is frequently used to produce intricate, high-precision parts in sectors such as aerospace, automotive, and healthcare. The SLS technique uses a laser beam to sinter powdered polymer materials, such as nylon, creating layers of powdered materials that are then built upon each other to create intricate 3D parts. An electron gun is used in place of the laser in EBM, which is similar to SLM. Reduced material waste, better resolution, shorter product development times, reduced need for machining fixtures, faster prototyping, and better low-volume production are some of the benefits of the PBF process. Lengthy and slow print time, frequent post-processing, thermal distortion, poor surface quality, and high cost due to higher powder consumption are some of the drawbacks of the process.

1.2.1.2 Wire Arc Additive Manufacturing

WAAM uses an electric arc as a heat source to melt a metallic wire, depositing it layer by layer to build a 3D object. Traditional welding methods like metal inert gas, tungsten inert gas, or plasma arc welding are the foundation of the procedure, which is enhanced with robotic systems for control and precision. Additionally, it can be used to maintain and replace broken parts and components. The direct cost of metal wire utilized in the WAAM process is typically about 10% of the weight of the metal powder [29]. In comparison to traditional processes, WAAM is primarily used for the production of large-scale, high-strength metal components, resulting in less material waste and shorter lead times. It is used to produce structural and repair parts for the automotive, marine, and aerospace sectors.

WAAM offers advantages for 3D printing thin-walled structures, such as high deposition rates, cost-effective material utilization, and compatibility with a wide range of metals. It is ideal for large-scale applications, producing parts with good mechanical strength and minimal material waste. However, challenges include high heat input, which can cause warping or distortion, residual stresses, poor surface finish requiring post-processing, and difficulties in maintaining consistent wall thickness [35,36]. Additionally, WAAM struggles with dimensional accuracy and handling intricate geometries, making it less suitable for highly detailed thin-walled designs.

1.2.1.3 Fused Deposition Modeling

FDM is a widely used 3D printing process that builds objects layer by layer using thermoplastic materials. The process begins with a 3D model created in CAD software, which is sliced into horizontal layers to generate a G-code file containing printer instructions. A thermoplastic filament, such as polylactic acid (PLA), nylon, polycarbonate (PC), polyetherketoneketone, and acrylonitrile butadiene styrene (ABS), is loaded into the printer, melted in the heated extruder, and extruded through a nozzle onto the build platform [32]. The printer deposits material sequentially, adhering each layer to the previous one until the object is fully formed. FDM is renowned for its affordability, ease of use, and versatility, making it a popular choice for rapid prototyping, educational purposes, and hobbyist projects. However, it has limitations, such as poor dimensional accuracy, visible layer lines, and reduced mechanical strength, particularly along the layer direction [31]. Thin-walled structures can also be prone to warping or delamination due to uneven cooling, and achieving high-quality prints often requires slower speeds and additional support.

1.2.1.4 Selective Laser Sintering

SLS is an AM process that builds parts layer by layer using a laser to sinter powdered material, typically polymers like nylon or composites [34]. The process begins with a CAD that is converted into a sliced digital file for the printer. A thin layer of powder is spread across the build platform, and the laser selectively heats and sinters (fuses) the powder particles together in specified areas according to the digital model [37]. The platform then lowers, and a fresh layer of powder is spread, repeating the process until the part is fully built. SLS does not require support structures because the unsintered powder surrounding the part provides natural support during the printing process. This enables the creation of complex geometries and intricate designs.

1.2.1.5 Stereolithography

Stereolithography (SLA) is the first developed AM process, and it is a high-precision AM process that creates parts layer by layer using a photopolymer resin cured by a focused ultraviolet (UV) laser. The process starts with a CAD model, which is converted into a sliced digital file to guide the printer. A thin layer of liquid resin is spread across the build platform inside a vat, and the UV laser selectively cures (solidifies) the resin in the areas corresponding to the model's design [38]. The platform then lowers by a small increment, and a new layer of resin is applied and cured, repeating this process until the entire part is completed. SLA is known for producing parts with exceptional accuracy, fine details, and smooth surface finishes, making it ideal for applications in prototyping, dentistry, jewelry, and engineering [37].

1.2.2 Matfrials for AM Process

The materials used in thin-walled structures play a critical role in determining their performance, durability, strength, weight, and suitability for specific applications. These structures are employed in various high-performance industries, such as aerospace, automotive, construction, and biomedical engineering, where materials must meet stringent requirements, including high strength-to-weight ratios, corrosion resistance, thermal stability, biocompatibility, and manufacturability. Commonly used materials are metals, polymers, ceramics, and composites. Table 1.1 presents the various AM thin-walled structures, along with the process, materials, and their corresponding thicknesses for thin-walled structures.

1.2.2.1 Metals

Metals and their alloys are the most prevalent materials for thin-walled structures due to their superior mechanical properties, versatility, and manufacturability. Aluminum alloys, for instance, are widely used in the aerospace and automotive industries for their lightweight, corrosion-resistant, and machinable properties. Titanium alloys, with their exceptional strength, biocompatibility, and resistance

TABLE 1.1
Different types of additively manufactured thin-walled structures

Material	Wall thickness	AM process used	Ref.
316L	0.25–0.35 mm	LPBF	[41]
IN718	0.6, 0.8, 10 mm	LPBF	[47]
IN718	1.5–10 mm	LMD	[42]
Inconel 625	1–1.2 mm	LPBF	[48]
$AlSi_{10}Mg$	0.1–10 mm	LPBF	[40]
$AlSi_{10}Mg$	0.5–5.0 mm	LPBF	[49]
Ti6Al4V	1–5 mm	EBM	[50, 51]
GRCop-42 alloy	0.7–2 mm	LPBF	[52]
PLA	1.2 mm	FDM	[33]
PLA, ABS	0.8 mm	Extrusion-based AM	[23]
Yttria-stabilized zirconia	2.55–3.33 mm	LENS	[26]
Al_2O_3	340–650 μm	SLA	[14, 53]
Al_2O_3	3 mm	SLA	[54]

to extreme environments, are frequently employed in aerospace components and biomedical implants [39]. Steel and stainless steel remain cost-effective options for applications that demand high durability and structural integrity. At the same time, magnesium alloys are favored for weight-critical designs in the automotive and consumer electronics industries. Emerging materials, such as high-entropy alloys (HEAs), are gaining traction due to their remarkable strength, corrosion resistance, and tailored properties, making them suitable for advanced engineering applications.

In the Al alloys category, Al-Si is a widely studied alloy for AM using metals. The $AlSi_{10}Mg$, with a wall thickness ranging from 0.1 mm to 10 mm, was fabricated by LPBF to study its microstructure [10,40]. The 316L steel, a type of austenitic stainless steel and precipitation-hardened steel, with a wall thickness ranging from 0.2 mm to 3 mm, was analyzed for its microstructure, manufactured by LPBF [41]. The Inconel 718 thin walls (1.5 mm) manufactured via LMD exhibited a coarsened microstructure, whereas the thicker wall thickness (10 mm) displayed a refined microstructure due to a strong temperature gradient [42]. Thin-walled single-crystal Ni-based superalloys are also useful in high-temperature and high-stress applications, such as turbine blades and aircraft vanes, due to their high thermal strength, toughness, creep resistance, and resistance to oxidation and corrosion [28]. In the AM of metal thin-walled structures, specialized alloys like GRCop-42 (Cu–4at.% Cr–2at.% Nb) have been explored for lightweight applications that require high thermal conductivity, such as heat exchangers and rocket engine liners [10].

HEAs, known for exceptional mechanical properties in extreme environments, are also promising for thin-walled structures. HEAs typically form single-phase solid solutions due to their high configurational entropy, with compositions that can be tailored for specific properties [43]. Studies on HEAs in thin-walled applications are limited. For example, LPBF-fabricated CoCrFeMnNi thin walls exhibit ribbon-like morphologies with long columnar grains growing epitaxially along the wall center, lower dislocation density, and discrete cell walls compared to bulk structures, owing to fewer thermal cycles [44]. Other HEAs, such as AlCoCrFeNi [45] and AlCuFeNiTi [46], have shown superior mechanical properties like high strength, fracture toughness, lower density, and thermal stability, making them promising candidates for thin-walled applications where both weight reduction and durability are crucial. In contrast, while widely used, conventional alloys such as stainless steels or titanium alloys often face limitations in maintaining their performance under extreme conditions or complex loading.

1.2.2.2 Polymers

Polymers are widely used in 3D printing for their versatility, ease of processing, and diverse mechanical properties [55]. Thin-walled polymer structures and composites are applied across industries like aerospace, automotive, medical, and architecture for rapid prototyping and low-volume manufacturing. Their lightweight yet strong nature makes them suitable for load-bearing applications [56]. Different AM techniques process various polymers, including thermoplastics, thermosets, elastomers, and polymer composites, using feedstocks like liquid polymers, powders, filaments, and films [57]. The most widely used thermoplastic polymer materials in FDM techniques include epoxy resin, PLA, polyamide (Nylon), ABS, and PC. Most frequently, polyamides (Nylon PA 11) and other thermoplastic polyurethane (TPU) are utilized in SLS and multijet printing. A variety of photo-active special polymers have been developed for the SLA process.

Techniques such as FDM [58], SLA [59], and SLS [38] are commonly used, with materials like thermoplastic, Nylon 11, Nylon 12, and TPU processed in powder form, and photopolymers used in liquid form. Recent advancements focus on polymer-based composites to enhance strength. Reinforcements, such as metals, ceramics, carbon, and natural fillers, improve mechanical, thermal, and electrical properties. Techniques like MM projection SLA enable the fabrication of carbon fiber-reinforced polymers (FRPCs), offering high stiffness, damping performance, and improved thermal conductivity [60]. An extensible AM method for producing FRPC composites, termed MM projection SLA, was introduced by Xu et al. [61]. High stiffness and damping performance were demonstrated experimentally by the developed two-phase micro lattice composites. Enhanced printing resolutions now allow the creation of complex interlocking designs and multi-directional composite preforms.

1.2.2.3 Ceramics

Advanced ceramics are widely used in thin-walled structures due to their thermal stability and hardness, making them ideal for high-temperature and wear-intensive applications such as turbine blades, heat exchangers, and biomedical implants [13]. AM has enabled the production of complex ceramic parts that are difficult or impossible to fabricate using conventional methods [62].

AM techniques for ceramics include slurry-based, powder-based, and bulk solid-based approaches, with standard methods such as inkjet printing, PBF, paste extrusion, and SLA. Inkjet printing offers high density and eliminates post-processing but requires careful control of particle size, rheology, and drying to ensure precision and avoid defects [63]. Other processes, such as direct ceramic SLA and extrusion-free forming, often require post-curing to achieve the desired properties [64]. Challenges in ceramic AM include brittleness, cracking, porosity, and poor layer bonding. Optimizing parameters such as particle composition, sintering conditions, and feedstock properties can improve mechanical strength and dimensional stability [65]. Recent advances include 3D printing of ceramic-graphene composites and barium titanate using specialized techniques, expanding the potential for high-performance ceramic applications.

1.2.2.4 Composites

Composite materials play a vital role in the development of thin-walled structures due to their customizable mechanical properties and lightweight characteristics. Among these, CFRPs [66] are notable for their exceptional strength-to-weight ratio and fatigue resistance, making them ideal for demanding applications such as aerospace fuselages, wind turbine blades, and racing car frames. Glass fiber-reinforced polymers, while more cost-effective, are suited for medium-performance uses like automotive panels and marine components [67]. Additionally, Kevlar and Aramid composites, renowned for their superior tensile strength and wear resistance, are widely utilized in high-impact applications, including bulletproof materials and aerospace panels.

Metal matrix composites (MMCs) offer outstanding mechanical properties, including a high strength-to-weight ratio, increased hardness, and corrosion resistance, making them superior to many conventional materials [68]. While traditional manufacturing methods can produce MMCs, they are often limited in creating complex geometries. AM has emerged as a viable alternative, enabling the fabrication of intricate MMC structures with enhanced properties. Commonly used matrix materials include aluminum (Al), magnesium (Mg), copper (Cu), titanium (Ti), nickel (Ni), and iron (Fe), with reinforcements such as aluminum oxide (Al_2O_3), silicon carbide (SiC), tungsten carbide (WC), and titanium carbide (TiC), among others [69].

1.3 CHALLENGES AND FUTURE SCOPE

The AM of thin-walled structures presents several significant challenges, primarily due to the geometry, material properties, and process-specific factors associated with thin sections. These challenges result in significant differences in the final geometrical accuracy, microstructure, and mechanical properties compared to bulk structures.

1.3.1 Heat Accumulation and Residual Stress

One of the most critical challenges is distortion and warping due to heat accumulation. Thin-walled components have a higher surface-area-to-volume ratio, leading to uneven thermal expansion and contraction during the manufacturing process. This results in residual stresses, which can accumulate due to uneven cooling rates, causing warping and dimensional inaccuracies [10]. In the DED process, residual stresses in compression can reach up to 100 MPa along both the print and build directions, and cumulative plastic strain reaches up to 40% near the substrate [70]. Residual stresses have been observed to reduce the effective elastic properties of lattice structures by approximately 6%–10% across three different material grades produced using the SLM process, and the residual stresses can reach values as high as 300–400 MPa due to rapid thermal cycles and steep temperature gradients [71]. Ali et al. reported that preheating SLM Ti-6Al-4V at 100°C results in 214 MPa residual stress, which drops to 61 MPa (71% reduction) at 370°C and further to 25 MPa (88.3% reduction) at 470°C [72]. These distortions not only affect the final part's shape but can also compromise its structural integrity and dimensional precision, making it difficult to achieve the required tolerance. Residual stress can reduce the structural stiffness of thin-walled components, making them prone to failure under mechanical loads. Post-process heat treatment can help relax these residual stresses, though achieving dimensional stability remains a significant challenge, especially in metal thin-walled structures.

1.3.2 Surface Roughness

Surface roughness is a critical concern in thin-walled structures, where surface defects such as porosity, waviness, and uneven layers dominate a larger proportion of the cross-sectional area due to the limited material volume. These surface defects degrade the mechanical properties and reduce the overall quality of the component [73]. Process parameter optimization is crucial for minimizing surface roughness; however, removing surface defects can inadvertently reduce wall thickness, further complicating the manufacturing process.

1.3.3 Process Control

Maintaining process control is another challenge due to the sensitivity of thin-walled structures to process parameters. Factors such as layer height, scan speed, and heat input play a crucial role in ensuring the quality of the build [74]. Variations in these parameters can lead to variations in wall thickness, poor surface quality, and reduced dimensional accuracy. Process control becomes even more critical as deviations in one parameter can amplify defects in thin sections. Specialized monitoring techniques can significantly enhance the quality of thin-walled structures during AM. Thermal imaging cameras (IR cameras, optical cameras, and X-ray imaging) can monitor temperature variations, preventing overheating or warping. At the same time, in-situ monitoring systems with feedback loops allow real-time adjustments to parameters like laser power and scan speed. Machine vision systems can track layer deposition and detect deviations in wall thickness or surface finish, while acoustic sensors monitor the sound during the process, identifying issues like over-melting [75]. Integrating these techniques improves process precision and enables early detection and correction of defects, enhancing the overall quality of thin-walled components.

1.3.4 Dimensional Deviations and Post-Process Machining

Dimensional deviations caused by heat accumulation and residual stress can be mitigated through post-process machining. However, machining very thin walls is challenging due to their low structural stiffness, leading to vibrations and large displacements of the cutter and workpiece [76]. This instability reduces the precision of machining processes, increasing lead times for fabricating thin-walled metal components. Hybrid additive and subtractive manufacturing approaches, where AM and machining are performed in a single workstation, offer a potential solution to enhance productivity and improve surface quality by addressing dimensional deviations during the printing process.

The future scope for thin-walled structures through AM is promising, driven by advancements in process control, material development, and hybrid manufacturing approaches. Improved process parameter optimization and in-situ monitoring will enable better control of thermomechanical phenomena, reducing heat accumulation, residual stresses, and surface defects to enhance geometrical accuracy and mechanical properties. Hybrid AM and subtractive manufacturing systems will enhance dimensional stability, productivity, and surface quality, addressing challenges such as warping and machining difficulties associated with thin-walled components. Material innovations, including high-performance alloys, functionally graded materials, and lightweight composites, will further enhance mechanical properties and structural stability. Additionally, AM's potential for sustainability, reduced material waste, and cost-effective production will drive broader adoption, while integration into existing manufacturing lines will enhance production efficiency and overall productivity.

REFERENCES

1. Arif, Zia Ullah, Muhammad Yasir Khalid, Reza Noroozi, Ali Sadeghianmaryan, Meisam Jalalvand, and Mokarram Hossain. "Recent Advances in 3d-Printed Polylactide and Polycaprolactone-Based Biomaterials for Tissue Engineering Applications." *International Journal of Biological Macromolecules* 218 (2022): 930–968.
2. Pesode, Pralhad, and Shivprakash Barve. "Additive Manufacturing of Metallic Biomaterials and Its Biocompatibility." *Materials Today: Proceedings* (2022). https://doi.org/10.1016/j.matpr.2022.11.248.
3. Pesode, Pralhad, and Shivprakash Barve. "Additive Manufacturing of Metallic Biomaterials: Sustainability Aspect, Opportunity, and Challenges." *Journal of Industrial and Production Engineering* 40, no. 6 (2023): 464–505.
4. Srivastava, Manu, Sandeep Rathee, Sachin Maheshwari, and T. K. Kundra. *Additive Manufacturing: Fundamentals and Advancements*. CRC Press, 2019.
5. Wang, Yuxuan, Yonghui Zhou, Lanying Lin, Jorge Corker, and Mizi Fan. "Overview of 3d Additive Manufacturing (Am) and Corresponding Am Composites." *Composites Part A: Applied Science and Manufacturing* 139 (2020): 106114.
6. Choudhari, Amit, James Elder, Manoj Mugale, Sanoj Karki, Venkata Bhuvaneswari Vukkum, Rajeev Kumar Gupta, and Tushar Borkar. "Additive Manufacturing of Aisi M2 Tool Steel by Binder Jetting (Bj): Investigation of Microstructural and Mechanical Properties." *Journal of Manufacturing Processes* 132 (2024): 686–711.
7. Liu, Shiwen, Shijie Chang, Haihong Zhu, Jie Yin, Tingting Wang, and Xiaoyan Zeng. "Effect of Substrate Material on the Molten Pool Characteristics in Selective Laser Melting of Thin Wall Parts." *The International Journal of Advanced Manufacturing Technology* 105 (2019): 3221–3231.
8 Du Plessis, Anton, Chris Broeckhoven, Ina Yadroitsava, Igor Yadroitsev, Clive H. Hands, Ravi Kunju, and Dhruv Bhate. "Beautiful and Functional: A Review of Biomimetic Design in Additive Manufacturing." *Additive Manufacturing* 27 (2019): 408–427.
9. Buchanan, Craig, and Leroy Gardner. "Metal 3d Printing in Construction: A Review of Methods, Research, Applications, Opportunities and Challenges." *Engineering Structures* 180 (2019): 332–348.
10. Viveganànthan, Priyanka, Shubo Gao, Weiming Ji, Haiyang Fan, Changjun Han, and Kun Zhou. "Recent Progress on 3d Printing of Lightweight Metal Thin-Walled Structures." *Advanced Materials* (2024): 2402130.
11. Bauer, Jens, Stefan Hengsbach, Iwiza Tesari, Ruth Schwaiger, and Oliver Kraft. "High-Strength Cellular Ceramic Composites with 3d Microarchitecture." *Proceedings of the National Academy of Sciences* 111, no. 7 (2014): 2453–2458.
12. Srivastava, Manu, and Sandeep Rathee. "Additive Manufacturing: Recent Trends, Applications and Future Outlooks." *Progress in Additive Manufacturing 7,* no. 2 (2022): 261–287.
13. Song, Haifeng, Jinqiu Tang, and Hongchang Zhang. "Additive Manufacturing and Performance Research of Aluminum-Based Ceramic Cores." *Journal of Physics: Conference Series,* 2720 (2024): 012025. doi:10.1088/1742-6596/2720/1/012025.
14. Kumar, M. Bhuvanesh, and P. Sathiya. "Methods and Materials for Additive Manufacturing: A Critical Review on Advancements and Challenges." *Thin-Walled Structures* 159 (2021): 107228.

15. Sonaye, Surendrasingh Y., Vijay K. Bokam, Akshay Saini, Vasudev V. Nayak, Lukasz Witek, Paulo G. Coelho, Sarit B. Bhaduri, Marco C. Bottino, and Prabaha Sikder. "Patient-Specific 3d Printed Poly-Ether-Ether-Ketone (Peek) Dental Implant System." *Journal of the Mechanical Behavior of Biomedical Materials* 136 (2022): 105510.
16. Pesode, Pralhad, and Shivprakash Barve. "Additive Manufacturing of Magnesium Alloys and Its Biocompatibility." *Bioprinting* 36 (2023): e00318.
17. Pesode, Pralhad, Shivprakash Barve, Sagar V. Wankhede, and Manoj Mugale. "Manufacturing Methods of Mg Alloys for Biomedical Applications." In Kumar A., Ansari A. (ed.) *Magnesium Alloys for Biomedical Applications*, 20–61. CRC Press, 2024.
18. Majeed, Arfan, Altaf Ahmed, Jingxiang Lv, Tao Peng, and Muhammad Muzamil. "A State-of-the-Art Review on Energy Consumption and Quality Characteristics in Metal Additive Manufacturing Processes." *Journal of the Brazilian Society of Mechanical Sciences and Engineering* 42 (2020): 1–25.
19. Ding, Junhao, Shuo Qu, Lei Zhang, Michael Yu Wang, and Xu Song. "Geometric Deviation and Compensation for Thin-Walled Shell Lattice Structures Fabricated by High Precision Laser Powder Bed Fusion." *Additive Manufacturing* 58 (2022): 103061.
20. Ni, Yujie, Ru Ji, Kaiwen Long, Ting Bu, Kejian Chen, and Songlin Zhuang. "A Review of 3d-Printed Sensors." *Applied Spectroscopy Reviews* 52, no. 7 (2017): 623–652.
21. Brown, Ben, Wes Everhart, and Joe Dinardo. "Characterization of Bulk to Thin Wall Mechanical Response Transition in Powder Bed Am." *Rapid Prototyping Journal* 22, no. 5 (2016): 801–809.
22. Chen, Changpeng, Zhongxu Xiao, Haihong Zhu, and Xiaoyan Zeng. "Deformation and Control Method of Thin-Walled Part During Laser Powder Bed Fusion of Ti–6al–4v Alloy." *The International Journal of Advanced Manufacturing Technology* 110 (2020): 3467–3478.
23. Bhatt, Prahar M., Rishi K. Malhan, Pradeep Rajendran, and Satyandra K. Gupta. "Building Free-Form Thin Shell Parts Using Supportless Extrusion-Based Additive Manufacturing." *Additive Manufacturing* 32 (2020): 101003.
24. Peng, Huaishu, Rundong Wu, Steve Marschner, and François Guimbretière. "On-the-Fly Print: Incremental Printing While Modelling." *CHI '16: Proceedings of the 2016 CHI Conference on Human Factors in Computing Systems* (2016): 887–896. doi.org/10.1145/2858036.285810.
25. Olivier, Djamila, Jose Antonio Travieso-Rodriguez, S. Borros, G. Reyes, and Ramón Jerez-Mesa. "Influence of Building Orientation on the Flexural Strength of Laminated Object Manufacturing Specimens." *Journal of Mechanical Science and Technology* 31 (2017): 133–139.
26. Fan, Zhiqi, Yitian Zhao, Mingyuan Lu, and Han Huang. "Yttria Stabilized Zirconia (YSZ) Thin Wall Structures Fabricated Using Laser Engineered Net Shaping (Lens)." *The International Journal of Advanced Manufacturing Technology* 105 (2019): 4491–4498.
27. Dejene, Naol Dessalegn, Hirpa G. Lemu, and Endalkachew Mosisa Gutema. "Critical Review of Comparative Study of Selective Laser Melting and Investment Casting for Thin-Walled Parts." *Materials* 16, no. 23 (2023): 7346.
28. Saleh Alghamdi, Saad, Sabu John, Namita Roy Choudhury, and Naba K. Dutta. "Additive Manufacturing of Polymer Materials: Progress, Promise and Challenges." *Polymers* 13, no. 5 (2021): 753.
29. Wickramasinghe, Sachini, Truong Do, and Phuong Tran. "Fdm-Based 3d Printing of Polymer and Associated Composite: A Review on Mechanical Properties, Defects and Treatments." *Polymers* 12, no. 7 (2020): 1529.
30. Wu, Bintao, Zengxi Pan, Donghong Ding, Dominic Cuiuri, Huijun Li, Jing Xu, and John Norrish. "A Review of the Wire Arc Additive Manufacturing of Metals: Properties, Defects and Quality Improvement." *Journal of Manufacturing Processes* 35 (2018): 127–139.

31. Shah, Abid, Rezo Aliyev, Henning Zeidler, and Stefan Krinke. "A Review of the Recent Developments and Challenges in Wire Arc Additive Manufacturing (Waam) Process." *Journal of Manufacturing and Materials Processing* 7, no. 3 (2023): 97.
32. Solomon, I. John, P. Sevvel, and Jjmtp Gunasekaran. "A Review on the Various Processing Parameters in Fdm." *Materials Today: Proceedings* 37 (2021): 509–514.
33. Khan, Imran, Imad Barsoum, Muhammad Abas, Ans Al Rashid, Muammer Koç, and Muhammad Tariq. "A Review of Extrusion-Based Additive Manufacturing of Multi-Materials-Based Polymeric Laminated Structures." *Composite Structures* 349–350 (2024): 118490.
34. Kafle, Abishek, Eric Luis, Raman Silwal, Houwen Matthew Pan, Pratisthit Lal Shrestha, and Anil Kumar Bastola. "3d/4d Printing of Polymers: Fused Deposition Modelling (Fdm), Selective Laser Sintering (Sls), and Stereolithography (Sla)." *Polymers* 13, no. 18 (2021): 3101.
35. Gueche, Yanis A., Noelia M. Sanchez-Ballester, Sylvain Cailleaux, Bernard Bataille, and Ian Soulairol. "Selective Laser Sintering (Sls), a New Chapter in the Production of Solid Oral Forms (Sofs) by 3d Printing." *Pharmaceutics* 13, no. 8 (2021): 1212.
36. Rodrigues, Tiago A., Valdemar Duarte, R. M. Miranda, Telmo G. Santos, and J. P. Oliveira. "Current Status and Perspectives on Wire and Arc Additive Manufacturing (Waam)." *Materials* 12, no. 7 (2019): 1121.
37. Pesode, Pralhad, and Shivprakash Barve. "Comparison and Performance of A, A+ B and B Titanium Alloys for Biomedical Applications." *Surface Review and Letters* 30, no. 12 (2023): 2330012.
38. Takata, Naoki, Hirohisa Kodaira, Asuka Suzuki, and Makoto Kobashi. "Size Dependence of Microstructure of ALSi10mg Alloy Fabricated by Selective Laser Melting." *Materials Characterization* 143 (2018): 18–26.
39. Wrobel, Rafal, Lorenzo Del Guidice, Poriya Scheel, Nerea Abando, Xavier Maeder, Michalis Vassiliou, Ehsan Hosseini, Ralph Spolenak, and Christian Leinenbach. "Influence of Wall Thickness on Microstructure and Mechanical Properties of Thin-Walled 316l Stainless Steel Produced by Laser Powder Bed Fusion." *Materials & Design* 238 (2024): 112652.
40. Li, Ying, Jaromír Dlouhý, Martina Koukolikova, Abhilash Kirana, Jaroslav Vavřík, and Jan Džugan. "Effect of Deposit Thickness on Microstructure and Mechanical Properties at Ambient and Elevated Temperatures for Inconel 718 Superalloy Fabricated by Directed Energy Deposition." *Journal of Alloys and Compounds* 908 (2022): 164723.
41. Sanchez, Salomé, Peter Smith, Zhengkai Xu, Gabriele Gaspard, Christopher J. Hyde, Wessel W. Wits, Ian A. Ashcroft, Hao Chen, and Adam T. Clare. "Powder Bed Fusion of Nickel-Based Superalloys: A Review." *International Journal of Machine Tools and Manufacture* 165 (2021): 103729.
42. Mugale, Manoj, Sanoj Karki, Amit Chaudhari, Satyavan Digole, Mayank Garg, Venkata A. S. Kandadai, Ganesh Walunj, Bharat K. Jasthi, and Tushar Borkar. "High Strength-Ductility Combination in Low-Density Dual Phase High Entropy Alloys." *Journal of Alloys and Compounds* 1010 (2024): 177762.
43. Zhang, Chengcheng, Kai Feng, Hiroyuki Kokawa, and Zhuguo Li. "On the Origin and Evolution of Cellular Structures in Cocrfemnni High Entropy Alloy Fabricated by Laser Powder Bed Fusion." *Materials Characterization* 196 (2023): 112586.
44. Munitz, Abraham, Shai Salhov, Shmuel Hayun, and Naum Frage. "Heat Treatment Impacts the Micro-Structure and Mechanical Properties of Alcocrfeni High Entropy Alloy." *Journal of Alloys and Compounds* 683 (2016): 221–230.
45. Mugale, Manoj, Mayank Garg, Ganesh Walunj, Venkata A. S. Kandadai, Bharat K. Jasthi, and Tushar Borkar. "Tweaking Alni Atomic Fraction to Enhance the Mechanical Properties of Low-Density Alcufeniti-Based High Entropy Alloys." *Journal of Alloys and Compounds* 1024 (2025): 180109.

46. Varney, Tracy Connor, Tylee Oldham, Md Imran Noor, and Paul F. Rottmann. "Quantifying the Microstructure and Mechanical Property Differences between Bulk and Thin-Wall Additively Manufactured Inconel 718." *Materialia* 31 (2023): 101867.
47. Banerjee, Arunima, Sara Messina, Matthew R. Begley, Edwin J. Schwalbach, Michael A. Groeber, William D. Musinski, Paul A. Shade et al. "The Mechanical Response of Additively Manufactured In625 Thin-Walled Structures." *Scripta Materialia* 205 (2021): 114188.
48. Zhang, Yingfeng, Arfan Majeed, Muhammad Muzamil, Jingxiang Lv, Tao Peng, and Vivek Patel. "Investigation for Macro Mechanical Behavior Explicitly for Thin-Walled Parts of Alsi10mg Alloy Using Selective Laser Melting Technique." *Journal of Manufacturing Processes* 66 (2021): 269–280.
49. Razavi, N., Brecht Van Hooreweder, and Filippo Berto. "Effect of Build Thickness and Geometry on Quasi-Static and Fatigue Behavior of Ti-6al-4v Produced by Electron Beam Melting." *Additive Manufacturing* 36 (2020): 101426.
50. Sepe, Raffaele, Stefania Franchitti, Rosario Borrelli, Francesco Di Caprio, Enrico Armentani, and Francesco Caputo. "Correlation between Real Geometry and Tensile Mechanical Behaviour for Ti6al4v Electron Beam Melted Thin Specimens." *Theoretical and Applied Fracture Mechanics* 107 (2020): 102519.
51. Sahoo, Subhadip, Mohammad M. Keleshteri, Jason R. Mayeur, and Kavan Hazeli. "Stress Localization Investigation of Additively Manufactured Grcop-42 Thin-Wall Structure." *Thin-Walled Structures* 201 (2024): 112022.
52. Dunaj, Paweł, Stefan Berczyński, Karol Miądlicki, Izabela Irska, and Beata Niesterowicz. "Increasing Damping of Thin-Walled Structures Using Additively Manufactured Vibration Eliminators." *Materials* 13, no. 9 (2020): 2125.
53. Halloran, John W. "Ceramic Stereolithography: Additive Manufacturing for Ceramics by Photopolymerization." *Annual Review of Materials Research* 46, no. 1 (2016): 19–40.
54. Zhang, Xueqin, Keqiang Zhang, Lu Zhang, Wenqing Wang, Ying Li, and Rujie He. "Additive Manufacturing of Cellular Ceramic Structures: From Structure to Structure–Function Integration." *Materials & Design* 215 (2022): 110470.
55. Bhalodi, Deep, Karan Zalavadiya, and Pavan Kumar Gurrala. "Influence of Temperature on Polymer Parts Manufactured by Fused Deposition Modeling Process." *Journal of the Brazilian Society of Mechanical Sciences and Engineering* 41 (2019): 1–11.
56. Yuan, Shangqin, Shaoying Li, Jihong Zhu, and Yunlong Tang. "Additive Manufacturing of Polymeric Composites from Material Processing to Structural Design." *Composites Part B: Engineering* 219 (2021): 108903.
57. Awasthi, Pratiksha, and Shib Shankar Banerjee. "Fused Deposition Modeling of Thermoplastic Elastomeric Materials: Challenges and Opportunities." *Additive Manufacturing* 46 (2021): 102177.
58. Lalegani Dezaki, Mohammadreza, Mohd Khairol Anuar Mohd Ariffin, and Saghi Hatami. "An Overview of Fused Deposition Modelling (Fdm): Research, Development and Process Optimisation." *Rapid Prototyping Journal* 27, no. 3 (2021): 562–582.
59. Huang, Jigang, Qin Qin, and Jie Wang. "A Review of Stereolithography: Processes and Systems." *Processes* 8, no. 9 (2020): 1138.
60. Mwania, Fredrick Mulinge, Maina Maringa, and Jacobus G. van der Walt. "A Review of the Techniques Used to Characterize Laser Sintering of Polymeric Powders for Use and Re-Use in Additive Manufacturing." *Manufacturing Review* 8 (2021): 14.
61. Benmesli, Samia, and Farid Riahi. "Dynamic Mechanical and Thermal Properties of a Chemically Modified Polypropylene/Natural Rubber Thermoplastic Elastomer Blend." *Polymer Testing* 36 (2014): 54–61.
62. Li, Neng, Shuai Huang, Guodong Zhang, Renyao Qin, Wei Liu, Huaping Xiong, Gongqi Shi, and Jon Blackburn. "Progress in Additive Manufacturing on New Materials: A Review." *Journal of Materials Science & Technology* 35, no. 2 (2019): 242–269.

63. Zhang, Feng, Zongan Li, Mengjia Xu, Shiyan Wang, Na Li, and Jiquan Yang. "A Review of 3d Printed Porous Ceramics." *Journal of the European Ceramic Society* 42, no. 8 (2022): 3351–3373.
64. Truxova, Veronika, Jiri Safka, Martin Seidl, Iaroslav Kovalenko, Lukas Volesky, and Michal Ackermann. "Ceramic 3d Printing: Comparison of Sla and Dlp Technologies." *MM Sci. J* 2020 (2020): 3905–3911.
65. Shen, Minhao, Wei Qin, Bohang Xing, Weiming Zhao, Shuyue Gao, Ying Sun, Ting Jiao, and Zhe Zhao. "Mechanical Properties of 3d Printed Ceramic Cellular Materials with Triply Periodic Minimal Surface Architectures." *Journal of the European Ceramic Society* 41, no. 2 (2021): 1481–1489.
66. Jia, Xiaohang, Junjie Luo, Quantian Luo, Qing Li, and Tong Pang. "Experimental Study on the Effects of Temperature on Mechanical Properties of 3d Printed Continuous Carbon Fiber Reinforced Polymer (Ccfrp) Composites." *Thin-Walled Structures* 205 (2024): 112465.
67. Raghava, P. Madhu, and G. Jaya Chandra Reddy. "3d-Printed Lightweight Sandwich Structures with Carbon-Fibre-Reinforced Polymer Skins and Glass-Fibre-Reinforced Polymer Core: A Comparative Study." *Smart Materials & Methods* (2024): 1–19.
68. Liu, Jianan, Ruiqing Ding, Jiaming Cao, Ke Zhan, Siquan Li, Bin Zhao, and Vincent Ji. "Recent Progress in Architecture Design of Nanocarbon-Reinforced Metal Matrix Composites and Their Properties: A Review." *Carbon* 228 (2024): 119382.
69. Behera, Malaya Prasad, Troy Dougherty, and Sarat Singamneni. "Conventional and Additive Manufacturing with Metal Matrix Composites: A Perspective." *Procedia Manufacturing* 30 (2019): 159–166.
70. Weisz-Patrault, Daniel, Pierre Margerit, and Andrei Constantinescu. "Residual Stresses in Thin Walled-Structures Manufactured by Directed Energy Deposition: In-Situ Measurements, Fast Thermo-Mechanical Simulation and Buckling." *Additive Manufacturing* 56 (2022): 102903.
71. Ahmed, Nissar, Imad Barsoum, and Rashid K. Abu Al-Rub. "Numerical Investigation of Residual Stresses in Thin-Walled Additively Manufactured Structures from Selective Laser Melting." *Heliyon* 9 (2023): 1–16.
72. Ali, Haider, Le Ma, Hassan Ghadbeigi, and Kamran Mumtaz. "In-Situ Residual Stress Reduction, Martensitic Decomposition and Mechanical Properties Enhancement through High Temperature Powder Bed Pre-Heating of Selective Laser Melted Ti6al4v." *Materials Science and Engineering: A* 695 (2017): 211–220.
73. Wu, Ziheng, Sneha Prabha Narra, and Anthony Rollett. "Exploring the Fabrication Limits of Thin-Wall Structures in a Laser Powder Bed Fusion Process." *The International Journal of Advanced Manufacturing Technology* 110, no. 1 (2020): 191–207.
74. Fang, Qihang, Gang Xiong, MengChu Zhou, Tariku Sinshaw Tamir, Chao-Bo Yan, Huaiyu Wu, Zhen Shen, and Fei-Yue Wang. "Process Monitoring, Diagnosis and Control of Additive Manufacturing." *IEEE Transactions on Automation Science and Engineering* 21, no. 1 (2022): 1041–1067.
75. AbouelNour, Youssef, and Nikhil Gupta. "In-Situ Monitoring of Sub-Surface and Internal Defects in Additive Manufacturing: A Review." *Materials & Design* 222 (2022): 111063.
76. Romberg, Stian K., Abrian I. Abir, Christopher J. Hershey, Vlastimil Kunc, and Brett G. Compton. "Structural Stability of Thin Overhanging Walls During Material Extrusion Additive Manufacturing of Thermoset-Based Ink." *Additive Manufacturing* 53 (2022): 102677.

2 Challenges in Additive Manufacturing of Thin-Walled Structures and Their Applications

Shubhavardhan Ramadurga Narasimharaju, Andrea El Hassanin, Fabio Scherillo, Antonino Squillace, and Antonio Viscusi

2.1 INTRODUCTION

Thin-walled structures (TWSs) are generally characterised by their smaller wall thickness (no strict cut-off value for thin walls but ideally not exceeding 3–5 mm) relative to their dimensions [1]. TWSs are prevalent in advanced engineering applications, such as aerospace, automotive, biomedical, transportation, electrical and electronics, marine, and shipbuilding, due to their excellent strength-to-weight ratio and effective material utilisation [2]. With the fast-changing socio-economic and political landscape, there is a growing interest in proposing robust innovative solutions to design, develop, and implement new lightweight, recyclable material components (metals, ceramics, polymers, foam, and composites) and energy-efficient manufacturing methods [3–9]. Lightweight not only reduces fuel consumption but also contributes to lower carbon footprints [10]. It is well acknowledged that additive manufacturing (AM) processes offer a significant advantage in fabricating practically any geometry, including very elaborate hollow and TWSs with comparable microstructure and mechanical properties, which are utilised across various already-mentioned functional industrial sectors. The structural and sub-structural components of these critical applications are generally in the form of lightweight, TWSs, lattice structures, or hybrid structures, which tend to lose their stability over time [11]. Stability constraints are the key contributing factors in extremely critical applications working under severe environments, for example, rocket engine components, aircraft fuselage, nuclear fusion reactor components, etc. [11]. Similarly, it is important to comprehend the stability constraints of intricate components of thin-walled biomedical applications, such as scaffolds, engineered tissues, cranial

DOI: 10.1201/9781003571209-2

implants, cardiac stents, etc. [12]. Hence, it is paramount to have an in-depth knowledge of materials, design constraints, challenges, and limitations during AM fabrication of functional thin-walled structural applications [11,12]. This chapter aims to touch on the perspectives of the AM TWS challenges and their applications covered in the literature so far.

The suitable materials for fabricating AM TWSs are metals, metal alloys, polymers, and ceramics in the form of powders, wires, or sheets, depending on the functional requirements [13]. The AM techniques used to manufacture metallic TWSs are Binder Jetting (BJ), Laser Powder Bed Fusion (LPBF), Electron Beam Melting (EBM), Directed Energy Deposition (DED), and Wire Arc Additive Manufacturing (WAAM). For polymers, Stereolithography (SLA), Fused Deposition Modelling, Multi Jet Fusion, and PolyJet fusion. For ceramics, BJ, SLA, and laser-based slurry deposition [11–13]. Most of AM processes use powder-form feedstock materials and utilise laser sources, electron beams, plasma beams, and electric arcs as heating sources to effectively manufacture TWSs. However, the laser-based and electron beam-based heating sources are the go-to AM techniques if one is looking for finer and better resolution [11]. The powder feedstocks dominate the AM market, but the wire feedstocks can also be easily manufactured; the resolution of wire-based feedstock parts is comparatively lower [12, 13]. The common challenges or limitations of metal-based AM TWSs are: (i) non-uniform phase change leading to internal stresses; (ii) formation of inevitable metallurgical defects and asperities, such as balling, porosities, spatters, lack of fusion holes, voids, residual stresses, thermal cracks, un-melted or partially melted powders, delamination, denudation, oxidation, etc. causing buckling, warping and distortion and resulting in poor surface finish; (iii) size constraint in the case of LPBF and EBM techniques, whereas poor resolution in the case of DED process; (iv) difficult to control heat source velocity and scanning or deposition pattern; and (v) difficult to remove support structures in support mandatory metal AM processes [14–21]. Challenges and limitations of ceramics and polymers are: (i) the refractive index of ceramics causes scattering, resulting in poor bonding and cohesion between the successive layers, lower toughness, cracking, and poor reproducibility, and (ii) the dampness effect, non-uniform energy absorption during sintering, etc. [22–24].

2.2 SPECIFIC CHALLENGES OF AM THIN-WALLED STRUCTURES

Although many reliable AM techniques like LPBF, EBM, and laser metal deposition (LMD) have shown signs of promise to overcome the critical issues associated with fabricating AM TWSs, the real challenges persist with the selection of the right material-process-system technology combination to achieve high productivity with precision from the wide variety of materials and processes [25]. TWSs can be efficiently manufactured by AM techniques but are inherently susceptible to a range of stability issues. It is important to address these challenges to avoid premature structural failure. The stability issues, including buckling behaviour, inadequate load resistance, and various other load-sensitive failure mechanisms, are shown in Figure 2.1 [3].

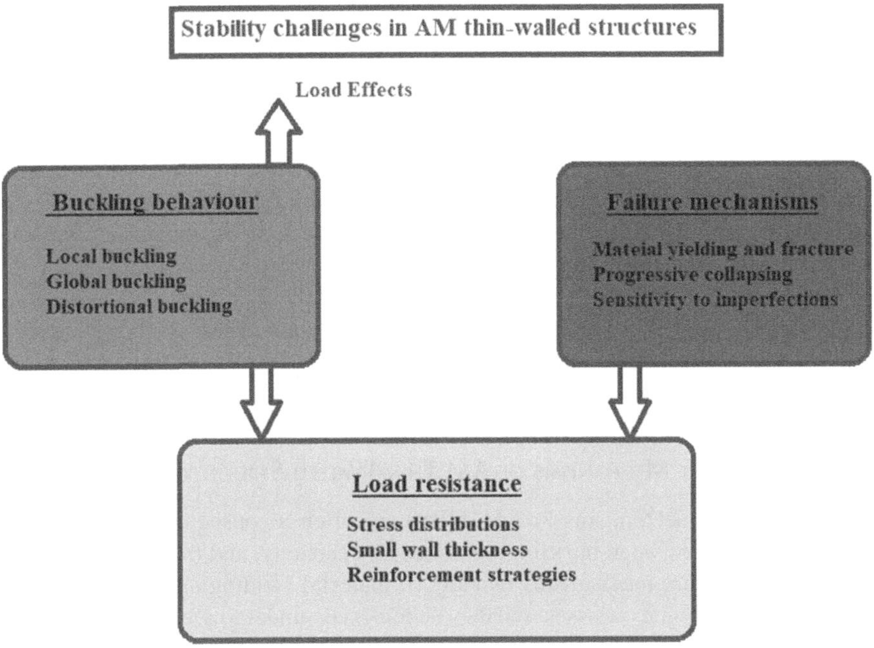

FIGURE 2.1 Schematic illustration of stability challenges in AM thin-walled structures. AM, Additive manufacturing.

2.2.1 Buckling Behaviour of AM Thin-Walled Structures

Buckling is the primary and most critical stability issue in AM TWSs. Due to their reduced wall thickness and slender cross-sections, the AM TWSs are susceptible to buckling under compressive stresses, even when their material strength is not exceeded.

The buckling of AM TWSs can be local, global buckling, or distortional buckling. Local buckling occurs in an individual plate component or its elements when their local stresses exceed the critical buckling stress [26,27]. The local buckling is often associated with residual stresses, defects, asperities, and other component imperfections [26,27]. As the name indicates, global buckling involves buckling of a whole component or the entire structural component or system due to overall instability under axial compression loads [28]. Slenderness, boundary conditions, and load eccentricity contribute to the global buckling [28]. Finally, the distortional buckling takes place when the cross-section of a TWS shape distorts, which could be due to the combination of both local effects and global instability [26,29]. The buckling behaviour can be mitigated by adopting the following measures: reinforcement, stiffening elements, optimised cross-sectional design, support enhancements, precise fabrication, tailored stiffeners, composite material usage, improved detailing, etc. [26–29].

2.2.2 Load Resistance of AM Thin-Walled Structures

The load resistance governs the safe operation of TWSs. Understandably, the smaller wall thickness of TWSs combined with complex thermophysical phenomena during AM may lead to uneven load distribution, resulting in localised overstressing and premature failure [3]. Employing artificial neural networks and machine learning tools, and advanced computational modelling, Finite Element Analysis (FEA), Computational Fluid Dynamics (CFD) for load resistance prediction models, validating them with experiments, has enabled envisioning of stress concentrations and the development of tailored design modifications, eventually improving the load-bearing capacities of AM TWSs [30,31]. The uniform distribution of load-carrying capacity and the overall structural stability can be enhanced by adopting advanced topology optimisation techniques and material grading [30,31].

2.2.3 Fracture Mechanisms of AM Thin-Walled Structures

The key failure mechanisms in AM TWSs are often a consequence of the intertwining relation between buckling, load-bearing capacity, and the material behaviour. These fracture mechanisms include: (i) material yielding and fracture: under the influence of higher stresses, the ductile materials undergo yielding and eventually resulting in fracture, whereas the brittle materials experience sudden failure before yielding; (ii) progressive collapsing: as soon as the buckling starts, this can rapidly propagate causing a chain reaction of catastrophic failure or collapsing; and finally, (iii) sensitivity to imperfection, the inevitable AM process-induced defects and asperities cause minor imperfections that adversely affect the buckling resistance, making thin-walled AM structures more sensitive and vulnerable [32,33]. Advancements in computational, Artificial Neural Network (ANN), Machine Learning (ML), topology optimisation techniques, and experimental methods are necessary for understanding these fracture mechanisms. The knowledge gained from these advanced tools is crucial for tailoring strategies to mitigate failure modes, such as improved material formulations, effective manufacturing practices with precision, and the integration of additional higher stability and stiffening features.

2.3 APPLICATIONS OF AM THIN-WALLED STRUCTURES

The applications of AM TWSs comprise metals, polymers, composites, etc. However, the metal AM TWSs are most prominent across various industrial sectors, namely aerospace, automotive, biomedical, energy, transportation, electrical and electronics, shipbuilding, marine, offshore, and wind turbine applications [1,2]. Utilising the benefits offered by AM in these industrial applications emphasises reduced material consumption, less material wastage, recycling, customised or tailored design, fabrication of structural and sub-structural components, reduced lead times, integration of complex TWS and lattice structure features, and repairing of the damaged functional components [2]. The applications of all the above-mentioned industrial sectors are enumerated in the following subsections.

2.3.1 Biomedical Applications

AM has shown greater promise and feasibility in the fabrication of customised, patient-specific implants, including cranial implants, collarbone implants, dental implants, knee and hip implants, as well as physician-specific complex-geometry surgical tools and instruments, as illustrated in Figure 2.2 [34,35]. Moreover, the ability of AM for innovative design of TWSs, and lattice-type and hybrid porous structures opens the door for fabrication of newer types of cardiovascular stents, biodegradable scaffolds, and spinal and various orthopaedic implants with tailored or optimised microstructural, surface finish and mechanical properties [34,36]. All these innovative new design TWSs are shown in Figure 2.3 [34,36]. The tailored or optimised thin-walled AM implants possess interconnected porosities which help in the easy

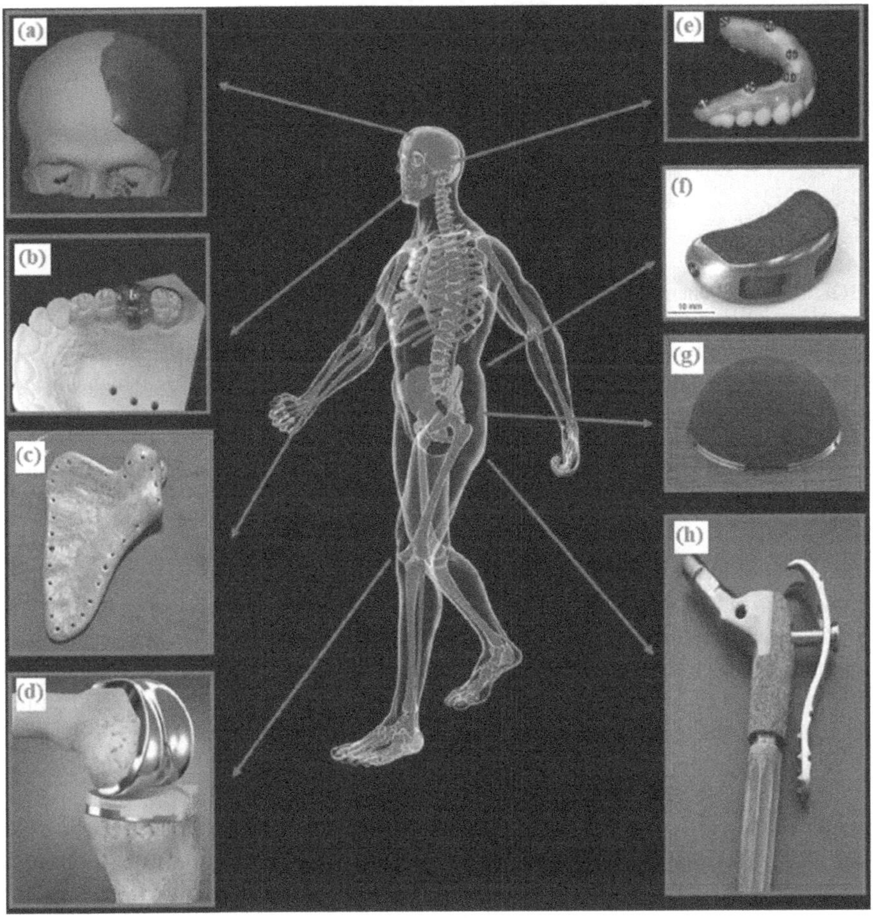

FIGURE 2.2 Schematic illustration of metal AM biomedical applications: (a) cranial prosthesis, (b) surgical guide, (c) scapula prosthesis, (d) knee prosthesis, (e) dental implants, (f) interbody fusion cage, (g) acetabular cup, and (h) hip prosthesis [34,35]. AM, Additive manufacturing.

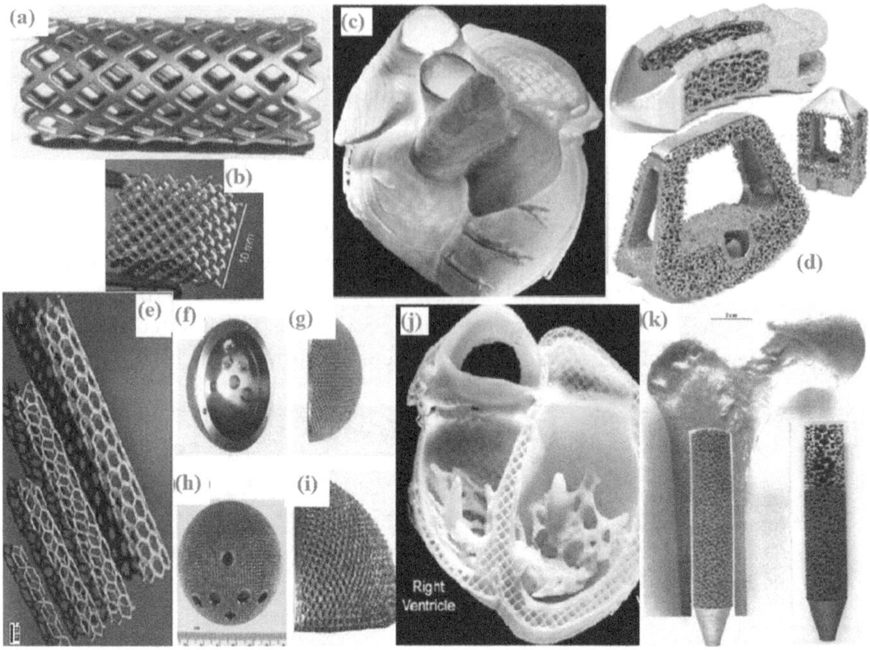

FIGURE 2.3 Schematic illustration of (a) cylindrical mesh spinal implant, (b) biodegradable scaffold with interconnected pores, (c) collagen printed heart, (d) typical AM porous spinal implant, (e) cardiovascular stents, (f)–(j) a metal AM-fabricated acetabular cup with porous mesh structure (outside) enabling ingrowth: (f), (g) lateral views; (h), (i) higher magnification of the outer porous mesh structure, (j) sectional view showing the intricate structures inside the heart, and (k) metal AM-fabricated intramedullary rod insert [11,34,35,36]. AM, Additive manufacturing.

fluid flow for nutrients, waste transportation, in-cell growth, osseointegration to provide better biocompatibility and load-bearing capability, bone-bonding ability, and finally in eliminating the need for autografting [35,36]. The surface properties of biomedical implants, including dental implants, are fundamental to the adhesion and differentiation of osteoblasts at the beginning of the osteointegration [37]. The textured or rougher surface induced by diverse metallurgical defects and surface asperities assists in bone-to-implant contact [37].

The surface roughness properties are crucial in determining the functional performance of the various biomedical implants [34]. The thorough extraction of the existing literature revealed the promising roughness values (1–100 μm) that maximise the bone cells' behaviour in the healing of the tissue around the implant [38,39]. The widely available dental implant system's bone tissues can adapt to surface roughness properties within this range, and tailoring or modifying the surface topography of these implants could greatly improve their primary stability and osseointegration [40]. However, the ideal surface texture or roughness properties for primary stability and osseointegration of biomedical implants remained a challenge [34].

Most functional osseointegrated implants require a surface roughness of 0.5–1.5 μm; with the current state-of-the-art of AM, it is non-viable to achieve this extent of surface finish [34–40]. However, some of the recent research revealed that these implants' biocompatibility and load-bearing capabilities could be enhanced by designing/modifying and/or tailoring the surface roughness to greater than 4 μm (up to 100 μm in dental implants) [37–40]. The external surface quality of acetabular bone implants is rougher from the outside and smoother from the inside, whereas the tibial augments possess a rougher internal surface and smoother outside [34]. Therefore, it is crucial to study and understand the holistic structural (both external and internal) surface properties to complement the existing microstructural and mechanical property studies applicable for all industrial sector applications [34]. There is a decent amount of research available on studying or optimising the microstructure and mechanical properties of AM TWSs used in diverse industrial sectors [10–20]. However, very limited research is focused on studying the surface roughness properties correlating with the functional performances of AM components, in particular AM TWSs. The complete understanding of microstructure, surface quality, or roughness influencing the final mechanical properties, which determines the functional performance of AM components, including AM TWSs, is necessary and must be addressed at the earliest opportunity.

It is important to gain complete insights into various AM surface topographies to optimise or tailor the desired surface quality [1–3]. The available literature completely lacks characterisation, correlation, and quantification of the diverse metallurgical defects and asperities that emerge on various metal AM surfaces [2]. Also, the existing research is limited to surface roughness height parameters (Ra/Sa), which give only decent descriptions of vertical characteristics of the surface deviation. However, it is highly unlikely to describe the surface after two processes (for example, a plateau-honed cylinder surface) [34]. The basic knowledge of roughness characterisation of Ra/Sa or a finite number of parameters will not provide detailed information to fully understand the variation of the surface roughness/texture and the suitable metrics to represent the complex metal AM surface topographies. Ra/Sa are insensitive to differentiate peaks, valleys, and the spacing of the various texture features. Also, Ra/Sa may be misleading on many surfaces with grossly different spatial and height symmetry features (for example, milled vs honed) that may have the same Ra/Sa but function quite differently. Ra/Sa fails to provide the general relation between the wavelength and amplitude properties of the surface, or any changes that occur either in amplitude or spacing [34]. Hence, with mere Ra/Sa information, it is non-viable to optimise or tailor the surface quality to meet the desired requirements for product design benefits and functional performance evaluation [34].

2.3.2 Aerospace Applications

AM is perceived as a novel production method that opened new opportunities for the complex design of multifunctional lightweight porous lattice structures of wind tunnels, automotive parts, brackets, conformal cooling channels, thrust chambers, combustion chambers, fuel tanks, injection heads, power heads, impellers, compressors, turbine blades, heat exchangers, and thermal devices of space applications, including TWSs of rocket engine components from the functional, technical, and economical points of view, as presented in Figure 2.4 [41–56]. Cooling (or regenerative cooling) is essential for maintaining the structural integrity of various aerospace applications comprising high-performing rocket engines, which are continuously exposed to higher thermal and environmental conditions [45]. Regenerative cooling (coolant flows through passages around the chamber wall while dissipating heat from the wall) can control the wall temperatures and balance the structural margins of the design and material [45]. Several case studies of metal AM TWSs are available online. Norsk Titanium first additively manufactured the Ti6Al4V thin-walled structural component for the Boeing 787 Dreamliner, which was first approved by the

FIGURE 2.4 Schematic illustration of advanced metal AM thin-walled structures of a rocket engine in aerospace applications [41–43,46,48–52]. AM, Additive manufacturing.

Federal Aviation Administration in 2017 [46]. LPBF, EBM, and WAAM are widely used AM techniques to fabricate metallic thin-walled structural components [2].

As a case in biomedical applications, the surface roughness properties are key to the functional performance of aerospace applications. A certain (higher) surface roughness induced by intentional porosities adds more benefits for extracting heat from cooling and maintaining the structural integrity of rocket engines and other aerospace-related applications [21]. In the case of lubricated fretting and Marines' shoes, the rougher surface helps in more oil retention, resulting in smaller wear and friction as compared to a smoother surface [21].

The other metal AM functional TWS applications are shown in Figure 2.5 [41,53–56]. Utilising an AM ability of infinite design and fabrication of intricate parts resulted in producing a lightweight structural component (NASA's Perseverance rover) with uniform stress distribution, which also exceeded the minimum necessary stiffness requirement. The weight was reduced by over 40% and the stiffness by more than 30% as compared to the conventional fabricated component [41,57]. This perseverance roller consisted of fabricating five additional components used in a planetary instrument for X-ray lithochemistry, in which two pieces are Ti shells, accounting for the frame, and two hollow, thin supporting struts to secure the shell to the arm ends [41,57].

The Welding Institute fabricated a lightweight Inconel 718 helicopter combustion casing using a five-axis LMD AM process. This allowed the combustion casing printing of overhangs without using any support structures, quickly within 7.5 hours [58]. It is well acknowledged in the AM fraternity that the DED process poses a standalone advantage of repairing and restoring damaged or broken functional parts rather than building an entire part from scratch [59]. The TWSs are repaired using the

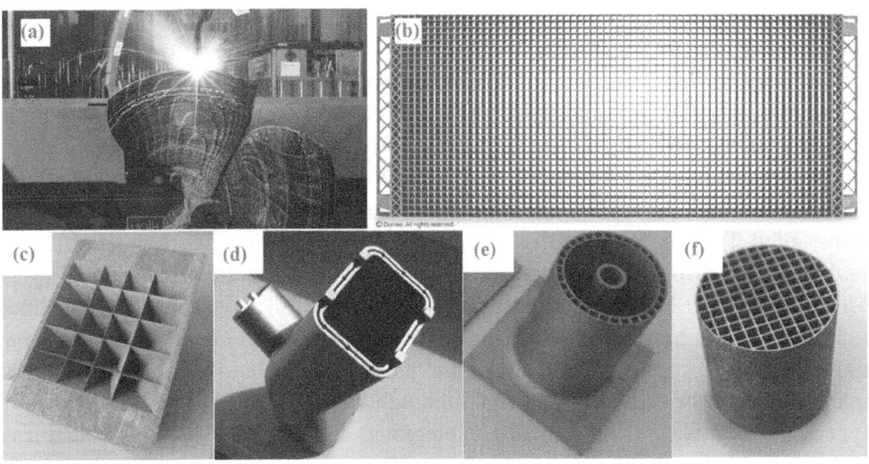

FIGURE 2.5 Schematic illustration of metal AM thin-walled structures in other applications: (a) WAAMpeller, (b) anti-scatter grids of XCT scanners, (c) lofthole collimator, nuclear fusion reactors blanket components: (d) flow channel inserts, (e) helium-cooled pebble bed fuel pin, and (f) stainless-steel honeycomb monolith support for a Ni-based catalyst [41,47–50]. AM, Additive manufacturing.

DED technique. For example, the eroded aerofoils of a blisk fabricated from AM355 steel experienced erosion during service. This aerofoil was repaired and restored by Optomec Laser Metal Deposition (LMD) DED with a more wear-resistant Stellite 21 alloy, as shown in Figure 2.4d. The restoration successfully passed the rigorous 60,000 RPM proof test and a 50,000-cycle low-cycle fatigue test, displaying excellent erosion, mechanical, and metallurgical properties [60].

The metal AM thin-walled structural components have shown greater promise for use as functional components in aerospace applications [41–45]. However, more opportunities present higher challenges. Some of the challenges are that there is decent literature available on studying the mechanical performances under static loading conditions [5,6,41–45], whereas the research on the surface quality, mechanical properties under dynamic loading conditions (fatigue properties), and creep behaviour is extremely limited [4]. The aerospace structural components experience severe demanding environments: high pressure, high temperature, highly corrosive, and cryogenic temperatures [41–45]. Hence, it is paramount to address the fatigue and creep behaviour of AM TWSs [10]. Similarly, the AM-process-induced inevitable metallurgical defects (distortion, residual stress, dimensional inaccuracy, and porosities), surface asperities generating anisotropic microstructures, and a poor surface finish determine the mechanical properties of TWSs [1,10,61]. The correlation between these defects, asperities, and surface finish that define the static and dynamic mechanical properties requires a complete understanding to meet the necessary specifications of industrial standards [61].

2.3.3 Automotive Applications

The primary concern in the automotive sector is to reduce fuel consumption and carbon emissions [1,10]. Like aerospace applications, lightweight, thin-walled structural components in automotive industrial applications consume less fuel and therefore emit a lower carbon footprint [10]. It is well acknowledged that AM is no longer just a prototyping technique; instead, it has evolved into producing customised, lightweight, complex end-use parts at reduced lead times [61]. Automotive is no different, which has utilised the benefits offered by AM by fabricating numerous functional automotive metal parts. For example, LPBF processing of the largest lightweight thin-walled Ti alloy brake callipers by Bugatti is shown in Figure 2.4f [50]. This brake calliper displayed a higher density of 99.7% and excellent mechanical properties (UTS 1250 MPa) under extreme temperatures [50]. Similarly, BMW printed lightweight aluminium alloy water pump wheels to withstand extreme operating conditions in racing cars [62]. Ford holds the place for printing the largest metal AM manifold [63], whereas Porsche printed the lightweight electric-drive housing prototype with optimised lattice structures, integrated cooling channels, higher stiffness, and reduced lead times, as shown in Figure 2.4g [64]. General Motors adopted the advanced generative design approach to design and AM-fabricate the strong (20%), lightweight (40%) complex stainless-steel bracket [65]. Some of the other lightweight automotive components are the car suspension double-wishbone components with different wall thicknesses fabricated by EBM. The wishbone part showed finer microstructure with excellent microhardness and was lighter by 38.6% compared to the conventional design and fabricated part [66].

Similar issues of AM TWS of aerospace applications persist with automotive industrial applications. The lack of complete understanding of process–structure–property interrelationships, especially related to dynamic behaviour conditions, is a major obstacle hindering the complete adoption of AM in the automotive industry. Also, the existing design procedures and standards are suitable for conventional mass production. Hence, the traditional design and production mindset of design engineers is hesitant to completely shift to exploring advanced lightweight AM TWSs in various functional automotive components.

2.3.4 Energy Applications

The impeller blades, turbines, and compressors are the ubiquitous TWSs in turbomachinery and energy generation. The unmatched ability of AM to design and create very complex, lightweight structural geometries of robust turbomachinery parts with better performance. For example, Siemens first produced the world's first turbomachinery parts, a gas burner for the SGT-700 gas turbine that operates at E.ON's combined cycle power plant in Germany [67]. Furthermore, Siemens produced many similar turbomachinery parts like burner heads, turbine blades, and compressors that can withstand working under extreme temperatures, higher stresses, and severe operating environments [67]. A very intriguing and true-to-scale miniature model of the Siemens SGT6-8000H gas turbine was manufactured by the Fraunhofer Institute for Manufacturing Technology and Advanced Materials and H+E-Produktentwicklung. This miniature model comprised only 68 parts created from steel, titanium, and aluminium, while the original component consisted of approximately 3000 parts [68]. Compared to aerospace, biomedical, and automotive AM TWS applications, the energy applications of AM TWSs are still in very early stages and need further development in the immediate future.

2.3.5 Thermal and Other Applications

Thermal applications of AM TWSs include heat exchangers, heat pipes, heat sinks, thermal energy storage, heat recovery, refrigeration cycles, and electronic equipment [69]. AM design and flexibility to print extremely critical geometries greatly improve the efficiency of these heat transfer device components. Also, AM printing of complex geometries increases the surface-area-to-volume ratio while avoiding the need to assemble multiple intricate parts [70]. The improved efficiency means that these critical thermal devices are made in compact sizes that are beneficial where lightweight and size restrictions are necessary. For example, the complex heat exchangers are based on heat tubes and plates [70]. Several innovative and advanced strategies are explored for improving the overall performance of AM thin-walled heat transfer devices through novel and generative designs [69]. These advanced tools mostly focused on modifying the intricate geometries, like permeable membrane microchannels, manifold microchannels, etc. For example, the (Triply Periodic Minimal Surface) TPMS lattice structure is shown in Figure 2.4i) [51]. The use of gradient structure has greater potential to improve the thermal protection property [52]; similarly, the organic-looking design for heat transfer devices also shows greater promise [71–73]. Figure 2.5 shows other promising applications of AM TWSs, such as ship propellers, wire-arc additive-manufactured WAAM peller, anti-scatter grids of X-ray computed tomography (XCT) scanners, lofthole collimators, nuclear fusion reactor blanket components: flow channel inserts, helium-cooled pebble bed fuel pins, and stainless-steel honeycomb monolith support for a Ni-based catalyst [4,53–56]. The WAAM part of the DED AM family is used to produce much larger parts than those fabricated by LPBF or LMD processes. The WAAM is limited to creating massive thin-walled AM structures, as its resolution is very high because the thickness of the printing layers is in the range of close to or higher than a millimetre, resulting in a poor surface quality compared to LPBF or EBM processes. The layer thickness produced by LPBF or EBM techniques is typically in micrometres; hence, you can expect to achieve a better surface finish [74]. The world's first class-approved ship propeller prototype (WAAMpeller) which was AM-printed in collaboration with Rotterdam Additive Manufacturing Lab and Damen Shipyards is shown in Figure 2.5a [75]. The three-blade and five-blade propellers were AM-printed and tested on a vessel, displaying equivalent results to the conventionally cast propellers [76]. Collimators are crucial thin-walled components that are extensively used in imaging instruments to minimise the background and unnecessary scattering of X-rays, gamma rays, and neutrons [77]. The collimators are extremely difficult to manufacture using conventional techniques like casting, drilling, milling, and electric discharge machining due to their highly complex geometries, and conventional processes are always labour-intensive while fabricating complex parts. Hence, AM is perceived to be an excellent alternative to fabricate collimators at shorter manufacturing times [54]. The AM thin-walled structured collimators are shown in Figure 2.5bc [54]. Notable other AM TWS applications can be found in nuclear fusion reactors, which are composed of flow channel inserts, dual coolant lithium blankets, cooling plates, and helium pebble bed fuel pins, as illustrated in Figure 2.5d,e [41,55,78]. The monoliths play an important role as supports for Ni/CeO_2-ZrO_2 catalysts. The bare monoliths displayed a very good conversion rate of CO_2 and CH_4 at high temperature (900oC) due to the presence of their inherent nickel content. For example, the stainless-steel honeycomb monolith structure has been successfully designed and produced using the LPBF process, which is schematically illustrated in Figure 2.5f [56].

REFERENCES

1. Sokolovskyi, M. V. et al. "Additive manufacturing of structural elements on a thinned-walled base: Challenges and difficulties." *The Paton Welding Journal* 11(2023): 46–52.
2. Zluhan, Bruno et al. "Design, defect analysis, compressive strength and surface texture characterization of Laser Powder Bed Fusion Processed Ti6Al4V Lattice structures." *Journal of Materials Research and Technology* 35 (2025): 2914–2933.
3. Azanaw, Girmay Mengesha. "Thin-walled structures in structural engineering: A comprehensive review of design innovations, stability challenges, and sustainable frontiers." *American Journal of Materials Synthesis and Processing* 10.1 (2025): 18–26.
4. Dubey, Rahul et al. "Energy absorption and dynamic behaviour of 6xxx series aluminium alloys: A review." *International Journal of Impact Engineering* 172 (2023): 104397.
5. Jobba, M., R. K. Mishra, and Marek Niewczas. "Flow stress and work-hardening behaviour of Al–Mg binary alloys." *International Journal of Plasticity* 65 (2015): 43–60.

6. Wang, Z., and M. X. Huang. "Optimising the strength-ductility-toughness combination in ultra-high strength quenching and partitioning steels by tailoring martensite matrix and retained austenite." *International Journal of Plasticity* 134 (2020): 102851.
7. Huang, Jinzhao et al. "A new partially-infused fiber reinforced thermoplastic composite for improving impact resistance." *International Journal of Impact Engineering* 168 (2022): 104293.
8. Du Bois, P. A. et al. "Material behaviour of polymers under impact loading." *International Journal of Impact Engineering* 32.5 (2006): 725–740.
9. Rahimidehgolan, Foad, Yue Liu, and William Altenhof. "Experimental, numerical and analytical investigations on the elevated strain rate compressive behavior of high-performance PES foam up to 200 s– 1." *International Journal of Impact Engineering* 161 (2022): 104088.
10. Yao, Ruyang et al. "On the crashworthiness of thin-walled multi-cell structures and materials: State of the art and prospects." *Thin-Walled Structures* 189 (2023): 110734.
11. Kumar, M. Bhuvanesh, and P. Sathiya. "Methods and materials for additive manufacturing: A critical review on advancements and challenges." *Thin-Walled Structures* 159 (2021): 107228.
12. Wu, Ziheng, Sneha Prabha Narra, and Anthony Rollett. "Exploring the fabrication limits of thin-wall structures in a laser powder bed fusion process." *The International Journal of Advanced Manufacturing Technology* 110.1 (2020): 191–207.
13. Noronha, Jordan et al. "Hollow-walled lattice materials by additive manufacturing: Design, manufacture, properties, applications and challenges." *Current Opinion in Solid State and Materials Science* 25.5 (2021): 100940.
14. Martin, John H., David S. Ashby, and Tobias A. Schaedler. "Thin-walled high temperature alloy structures fabricated from additively manufactured polymer templates." *Materials & Design* 120 (2017): 291–297.
15. Chen, Changpeng et al. "Deformation and control method of thin-walled part during laser powder bed fusion of Ti–6Al–4V alloy." *The International Journal of Advanced Manufacturing Technology* 110 (2020): 3467–3478.
16. Liang, Xuan et al. "Inherent strain homogenization for fast residual deformation simulation of thin-walled lattice support structures built by laser powder bed fusion additive manufacturing." *Additive Manufacturing* 32 (2020): 101091.
17. Algardh, Joakim Karlsson et al. "Thickness dependency of mechanical properties for thin-walled titanium parts manufactured by Electron Beam Melting (EBM)®." *Additive Manufacturing* 12 (2016): 45–50.
18. Demir, Ali Gökhan. "Micro laser metal wire deposition for additive manufacturing of thin-walled structures." *Optics and Lasers in Engineering* 100 (2018): 9–17.
19. Ding, Donghong et al. "A practical path planning methodology for wire and arc additive manufacturing of thin-walled structures." *Robotics and Computer-Integrated Manufacturing* 34 (2015): 8–19.
20. Li, Rong, Jun Xiong, and Yangyang Lei. "Investigation on thermal stress evolution induced by wire and arc additive manufacturing for circular thin-walled parts." *Journal of Manufacturing Processes* 40 (2019): 59–67.
21. Gockel, J., Fox, J., Beuth, J., and Hafley, R. "Integrated melt pool and microstructure control for Ti–6Al–4V thin wall additive manufacturing." *Materials Science and Technology*,31.8 (2015): 912–916.
22. Hu, Yingbin, and Weilong Cong. "A review on laser deposition-additive manufacturing of ceramics and ceramic reinforced metal matrix composites." *Ceramics International* 44.17 (2018): 20599–20612.
23. Halloran, John W. "Ceramic stereolithography: Additive manufacturing for ceramics by photopolymerization." *Annual Review of Materials Research* 46.1 (2016): 19–40.

24. Zocca, Andrea et al. "Additive manufacturing of SiSiC by layerwise slurry deposition and binder jetting (LSD-print)." *Journal of the European Ceramic Society* 39.13 (2019): 3527–3533.
25. Kruth, J.-P., Ben Vandenbroucke, Jonas Van Vaerenbergh, and Peter Mercelis. "Benchmarking of different SLS/SLM processes as rapid manufacturing techniques." In *International Conference Polymer & Moulds Innovations, PMI 2005*, p. 525. 2005.
26. Luo, Yangjun, and Junjie Zhan. "Linear buckling topology optimization of reinforced thin-walled structures considering uncertain geometrical imperfections." *Structural and Multidisciplinary Optimization* 62.6 (2020): 3367–3382.
27. Barbhuiya, Salim & Das, Bibhuti Bhusan. "Life Cycle Assessment of construction materials: Methodologies, applications and future directions for sustainable decision-making." *Case Studies in Construction Materials* 19 (2023): 1–33.
28. Ren, Rui et al. "Stiffness enhancement methods for thin-walled aircraft structures: A review." *Thin-Walled Structures* 201 (2024): 111995.
29. Olutola Tomiloba et al. "Sustainable Decision-Making for Thin-Walled Structures: A Comprehensive Review." *Asian Journal of Basic and Applied Sciences* 11.1 (2023): 399–403.
30. Li, Xiping et al. "Supportless 3D-printing of non-planar thin-walled structures with the multi-axis screw-extrusion additive manufacturing system." *Materials & Design* 240 (2024): 112860.
31. Mojtabaei, Seyed Mohammad et al. "Predicting the buckling behaviour of thin-walled structural elements using machine learning methods." *Thin-Walled Structures* 184 (2023): 110518.
32. Solazzi, Luigi. "Reliability evaluation of critical local buckling load on the thin walled cylindrical shell made of composite material." *Composite Structures* 284 (2022): 115163.
33. Gilbert, Benoit P. et al. "Thin-walled timber structures: An investigation." *Construction and Building Materials* 73 (2014): 311–319.
34. Narasimharaju, Shubhavardhan Ramadurga. "In the partial fulfilment of the requirements for the degree of doctor of philosophy in the department of mechanical engineering." *Doctora l dissertation*, University of Huddersfield (2023).
35. Ni, J. et al. "Three-dimensional printing of metals for biomedical applications." *Materials Today Bio* 3 (2019): 100024.
36. Meena, Vijay Kumar et al. "Additive manufacturing for metallic spinal implants: A systematic review." *Annals of 3D Printed Medicine* 3 (2021): 100021.
37. Matos, Geraldo Roberto Martins. "Surface roughness of dental implant and osseointegration." *Journal of Maxillofacial and Oral Surgery* 20 (2021): 1–4.
38. Elias, Carlos Nelson, and Luiz Meirelles. "Improving osseointegration of dental implants." *Expert Review of Medical Devices* 7.2 (2010): 241–256.
39. Albrektsson, Tomas, and Ann Wennerberg. "On osseointegration in relation to implant surfaces." *Clinical Implant Dentistry and Related Research* 21 (2019): 4–7.
40. Alla, R. Krishna et al. "Surface roughness of implants: A review." *Trends in Biomaterials and Artificial Organs* 25.3 (2011): 112–118.
41. Viveganathan, Priyanka et al. "Recent progress on 3D printing of lightweight metal thin-walled structures." *Advanced Materials* 36.49 (2024): 2402130.
42. Gradl, Paul. "Advanced and additive manufacturing technologies for liquid rocket engine components." In *Liquid Rocket Engines: Emerging Technologies in Liquid Propulsion*. 2020.
43. Gradl, Paul R., Sandy E. Greene, Christopher Protz, Brad Bullard, James Buzzell, Chance Garcia, Jessica Wood, Robin Osborne, James Hulka, and Kenneth G. Cooper. "Additive manufacturing of liquid rocket engine combustion devices: A summary of process developments and hot-fire testing results." In *2018 Joint Propulsion Conference*, p. 4625. 2018.

44. Ho, Jin Yao, Kai Choong Leong, and Teck Neng Wong. "Additively-manufactured metallic porous lattice heat exchangers for air-side heat transfer enhancement." *International Journal of Heat and Mass Transfer* 150 (2020): 119262.
45. Kerstens, Fabio, Angelo Cervone, and Paul Gradl. "End to end process evaluation for additively manufactured liquid rocket engine thrust chambers." *Acta Astronautica* 182 (2021): 454–465.
46. Norsk Titanium. https://www.norsktitanium.com/media/press, 2017.
47. EOS GmbH. https://www.eos.info, n.d.
48. Gradl, et al. *Additive manufacturing overview: Propulsion applications, design for and lessons learned*. No. M17-6434, 2017.
49. Gradl, Paul. https://ntrs.nasa.gov/, 2021.
50. Bugatti Communications. https://newsroom.bugatti.com/, 2018.
51. Khalil, Mohamad et al. "Forced convection heat transfer in heat sinks with topologies based on triply periodic minimal surfaces." *Case Studies in Thermal Engineering* 38 (2022): 102313.
52. Lin, Kaijie, Kaiming Hu, and Dongdong Gu. "Metallic integrated thermal protection structures inspired by the Norway spruce stem: Design, numerical simulation and selective laser melting fabrication." *Optics & Laser Technology* 115 (2019): 9–19.
53. Jorritsma, Joost and Albert-Michel Meyer. https://www.philips.com/, n.d.
54. Deprez, Karel et al. "Rapid additive manufacturing of MR compatible multipinhole collimators with selective laser melting of tungsten powder." *Medical physics* 40.1 (2013): 012501.
55. Koehly, C., H. Neuberger, and L. Bühler. "Fabrication of thin-walled fusion blanket components like flow channel inserts by selective laser melting." *Fusion Engineering and Design* 143 (2019): 171–179.
56. Agueniou, Fazia et al. "3D-printing of metallic honeycomb monoliths as a doorway to a new generation of catalytic devices: The Ni-based catalysts in methane dry reforming showcase." *Catalysis Communications* 148 (2021): 106181.
57. NASA Jet Propulsion Laboratory. https://www.jpl.nasa.gov/news, 2020.
58. The Welding Institute. https://www.twi-global.com/media-andevents, n.d.
59. Aprilia, Aprilia, Naien Wu, and Wei Zhou. "Repair and restoration of engineering components by laser directed energy deposition." *Materials Today: Proceedings* 70 (2022): 206–211.
60. Optomec. https://optomec.com, n.d.
61. Narasimharaju, Shubhavardhan Ramadurga et al. "A comprehensive review on laser powder bed fusion of steels: Processing, microstructure, defects and control methods, mechanical properties, current challenges and future trends." *Journal of Manufacturing Processes* 75 (2022): 375–414.
62. BMW Group. https://www.press.bmwgroup.com/global/article/, 2015.
63. Metal Additive Manufacturing Magazine. https://www.metal-am. com/, 2019.
64. Newsroom. P. https://newsroom.porsche.com, 2020.
65. Danon, Bill. https://adsknews.autodesk.com/, 2018.
66. Wang, Pan et al. "Experimental analysis of additively manufactured component and design guidelines for lightweight structures: A case study using electron beam melting." *Additive Manufacturing* 33 (2020): 101088.
67. Siemens. https://press.siemens.com/global/en, 2018.
68. Fraunhofer IFAM. https://www.ifam.fraunhofer.de/en/Press_Releases.html, 2019.
69. Al-khafaji, Osamah Raad Skheel, and Nizar F. O. Al-Muhsen. "Associated effect of perforation position and diameter on thermal performance of plane fin under natural convection environment." *Jurnal Teknologi (Sciences & Engineering)* 82.5 (2020):101–108.
70. Niknam,Seyed A., Mehdi Mortazavi, and Dongsheng Li. Additively manufactured heat exchangers: A review on opportunities and challenges. *The International Journal of Advanced Manufacturing Technology* 112.3 (2021): 601–618.

71. Lazarov, Boyan S. et al. "Experimental validation of additively manufactured optimized shapes for passive cooling." *Applied Energy* 226 (2018): 330–339.
72. Ho, Jin Yao et al. "An experimental investigation of a PCM-based heat sink enhanced with a topology-optimized tree-like structure." *Energy Conversion and Management* 245 (2021): 114608.
73. Iradukunda, Ange-Christian, Andres Vargas, David Huitink, and Danny Lohan. "Transient thermal performance using phase change material integrated topology optimized heat sinks." *Applied Thermal Engineering* 179 (2020): 115723.
74. Shubhavardhan Ramadurga Narasimharaju, et al. "Surface texture characterisation of longitudinal and latitudinal external and internal surfaces of laser powder bed fusion processed bespoke ball artefact". *Journal of Manufacturing Processes* 116 (2024): 120–148.
75. RAMLAB. https://www.ramlab.com/, 2017.
76. Naval News. https://www.navalnews.com/naval-news/, 2021.
77. Cramer, Corson L. et al. "Processing of complex-shaped collimators made via binder jet additive manufacturing of B4C and pressureless melt infiltration of Al." *Materials & Design* 180 (2019): 107956.
78. Lim, Bryan. et al. "Additively manufactured Haynes-282 monoliths containing thin wall struts of varying thicknesses." *Additive Manufacturing* 59 (2022): 103120.

3 The Additive Manufacturing Process for Thin-Walled Polymer, Ceramic, and Metal Structures

N. Harshavardhana, S. P. Sundar Singh Sivam, and M. Manjaiah

3.1 INTRODUCTION TO THIN-WALLED STRUCTURE USING ADDITIVE MANUFACTURING

Thin-walled structures are defined by their significantly smaller wall thickness relative to overall dimensions. They are often designed for lightweight applications while maintaining high structural integrity. Their unique geometry allows for reduced material usage without sacrificing mechanical strength, making them indispensable across a wide range of industries—including aerospace, automotive, biomedical, energy, and defense. These structures serve roles such as fuel tanks, heat exchangers, implants, turbine blades, and drone frames, offering optimized strength-to-weight ratios and enhanced functionality [1]. Table 3.1 shows the application of thin-walled structure applications across various industries [2].

While traditional manufacturing methods like machining, casting, metal forming, and drawing have long been used to produce thin-walled components, they often encounter limitations related to geometric complexity, material waste, and process inefficiencies. Furthermore, thin-walled geometries pose manufacturing challenges due to low stiffness, high aspect ratios, and susceptibility to warping, residual stress, and dimensional inaccuracies. Additive manufacturing (AM) offers a promising alternative by enabling the direct fabrication of complex, lightweight structures from digital models through a layer-by-layer approach. Unlike conventional techniques, AM allows for greater design freedom, minimal tooling, and the production of intricate geometries in a single step. Recent developments, particularly in metal AM,

DOI: 10.1201/9781003571209-3

TABLE 3.1
Thin-walled structure applications [1,2]

Industry	Applications
Aerospace	• Lightweight aircraft components (brackets, frames, structural panels, fuel tanks) • Rocket nozzles and satellite parts for weight reduction
Automotive	• Crash energy absorbers for safety • Lightweight body panels for fuel efficiency • Heat exchangers for improved thermal performance
Biomedical	• Customized orthopedic implants • Scaffolds for tissue engineering • Dental prosthetics with precise geometries
Energy	• Thin-walled heat exchangers for better thermal performance • Turbine components for enhanced efficiency • Lightweight enclosures for batteries and fuel cells
Defense & Military	• Lightweight drone frames for better maneuverability • Ammunition casings with high strength-to-weight ratio • Protective gear with improved durability
Electronics	• Compact enclosures for electronic devices • Heat dissipation systems for enhanced performance
Medical Devices	• Ergonomic surgical instruments • Lightweight diagnostic equipment
Marine & Offshore	• Lightweight hull components for better fuel efficiency • Corrosion-resistant piping systems for durability
Renewable Energy	• Solar panel frames for structural efficiency • Wind turbine blades for enhanced performance
Space Exploration	• Lightweight habitats for extraterrestrial missions • Satellite components optimized for transport and performance

have demonstrated the ability to fabricate thin-walled structures with a thickness below 100 μm, showcasing the technology's precision and scalability [3].

This chapter examines the role of AM in the fabrication of thin-walled structures across three primary material classes: metals, polymers, and ceramics. It explores the suitability of various AM techniques in addressing manufacturing challenges while balancing considerations such as material behavior, mechanical performance, geometric complexity, and post-processing requirements. By identifying the current limitations and advancements, the chapter aims to provide a comprehensive understanding of how AM technologies can be optimized for reliable and efficient production of thin-walled components in demanding applications.

3.2 DESIGN CONSIDERATIONS IN ADDITIVE MANUFACTURING

Design for AM, commonly referred to as 3D printing, represents a paradigm shift in modern manufacturing. This technology allows for the creation of complex three-dimensional objects by adding material layer by layer, directly from a digital design [4].

Figure 3.1 shows the schematic diagram for the design of AM [4,5]. Figure 3.1 outlines the end-to-end process involved in the AM of thin-walled components, highlighting five critical stages:

1. **Design** – Utilizes CAD/CAE software with advanced strategies like topological optimization and lattice structuring to achieve optimal strength-to-weight ratios.
2. **Verification** – Involves simulation and defect rectification to ensure design accuracy and model integrity before proceeding to fabrication.
3. **Slicing** – Converts 3D models into printable instructions, setting key printing parameters and selecting suitable materials tailored for thin-walled stability.
4. **Printing** – The component is additively manufactured using layer-by-layer deposition, with technologies such as wire arc additive manufacturing (WAAM), selective laser melting (SLM), or direct metal laser sintering (DMLS) commonly used for thin-walled structures.
5. **Finishing** – Encompasses the removal of support structures and application of surface or thermal treatments to enhance surface quality and mechanical properties [4].

AM offers unique opportunities for the precision fabrication of materials and structures that are difficult or even impossible to achieve using traditional methods. This is especially significant for applications requiring complex geometries, customized designs, or biocompatible architecture [6]. Figure 3.2 shows the various AM methods for producing thin-walled structures.

3.3 DESIGN OF LIGHTWEIGHT THIN-WALLED STRUCTURES IN ADDITIVE MANUFACTURING

Achieving lightweight thin-walled structures in AM involves advanced design strategies that optimize material usage while maintaining structural integrity. These structures are widely used in aerospace, automotive, and biomedical industries due to their excellent strength-to-weight ratios and energy absorption capabilities. Some of the methods for developing lightweight thin-walled structures are shown.

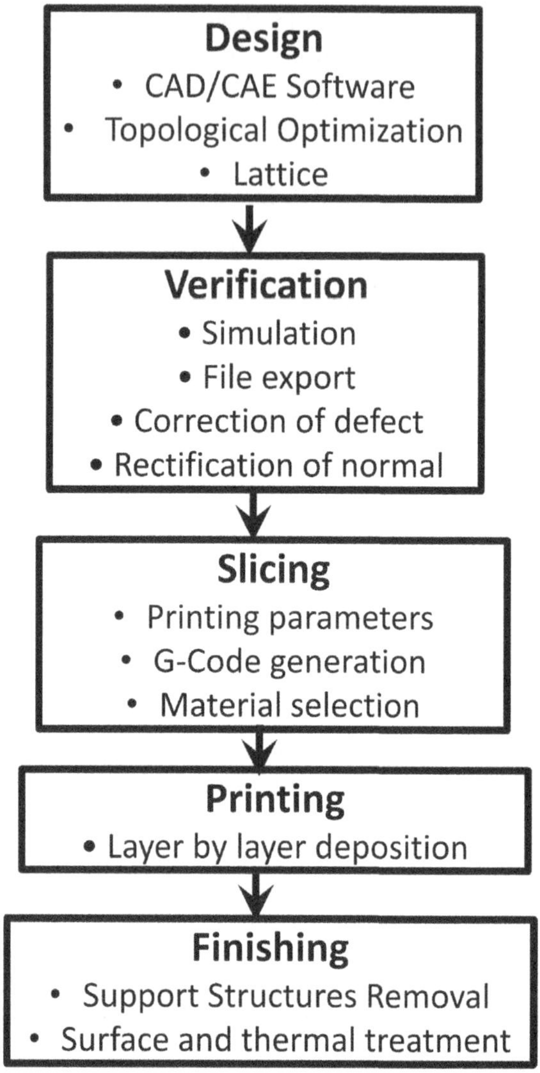

FIGURE 3.1 End-to-end process involved in the additive manufacturing (AM) of thin-walled components [4,5].

3.3.1 Lattice and Cellular Structures

Lattice structures consist of periodically arranged struts or beams forming an open-cell network. Due to their high porosity and substantial surface area-to-volume ratio, they are particularly advantageous for lightweight engineering applications. Their mechanical behavior is primarily influenced by parameters such as unit cell geometry, strut diameter, and the inherent properties of the material used [1].

These structures are known for their superior ability to distribute mechanical loads efficiently, making them ideal for scenarios where both strength and weight

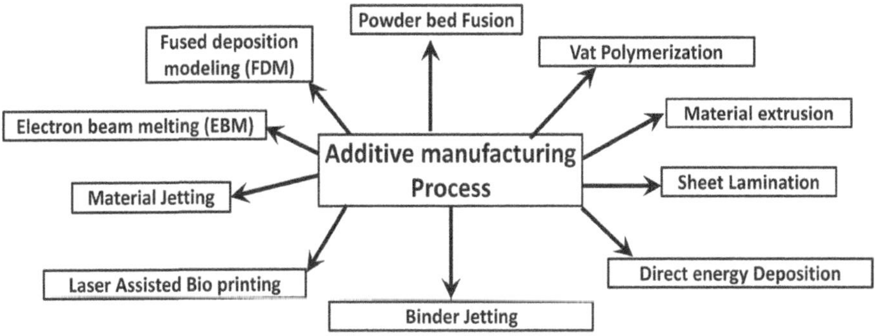

FIGURE 3.2 Schematic of various additive manufacturing methods for producing thin-walled structures [6].

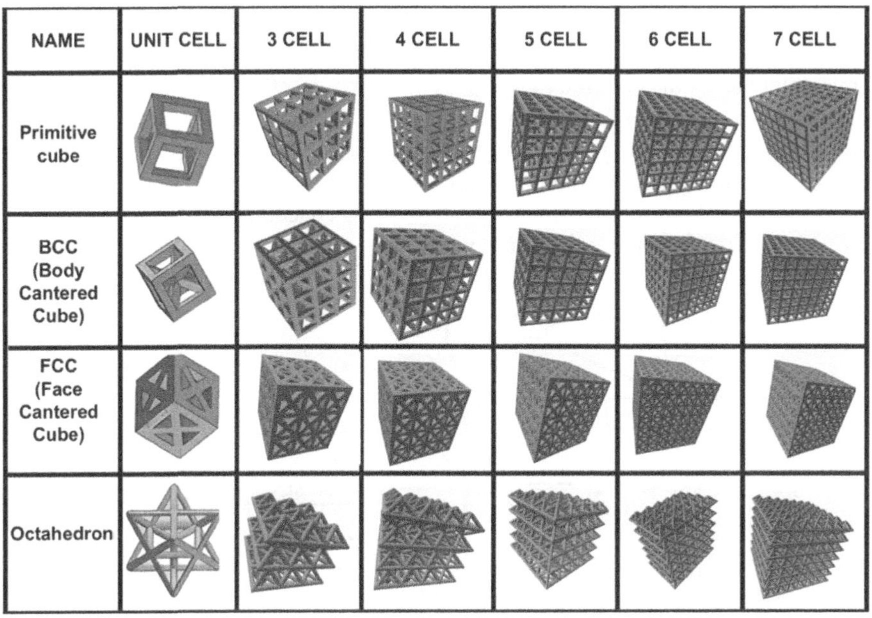

FIGURE 3.3 Lattice structures, including primitive cell, body-centered cube (BCC), face-centered cube (FCC), and octahedron lattice structure.

reduction are essential. Their structural efficiency can be further improved through the implementation of gradient density designs. In this approach, material is distributed non-uniformly—more in regions subject to higher stress and less in areas with lower stress. This tailored distribution enhances load-bearing capacity and overall structural performance. Four samples were fabricated for each type of lattice topology. Figure 3.3 represents the models of various lattice structures, including (a) primitive cell, (b) body-centered cubic (BCC), (c) face-centered cubic (FCC), and (d) octahedron [7].

On the other hand, cellular structures share similarities with lattice structures but typically feature closed-cell configurations, such as foams and honeycombs. These structures are well regarded for their energy absorption capabilities and thermal insulation properties, making them suitable for applications requiring impact resistance and heat management. The design process for cellular structures involves selecting appropriate cell shapes and sizes to balance mechanical performance and weight savings. For instance, honeycomb structures offer excellent strength and stiffness, whereas foam-based designs provide superior energy absorption. Figure 3.4 illustrates a schematic representation of various cellular structures and their configurations [4].

3.3.2 Topology Optimization

Topology optimization is a powerful computational method increasingly utilized for designing lightweight, thin-walled structures, particularly within the realm of AM. By intelligently redistributing material within a predefined design space, this approach enables the creation of components that are both structurally efficient and mechanically resilient. This section delves into the core concepts, simulation tools, and recent progress in topology optimization, with a focus on AI-driven generative design and multi-material solutions [3,8].

As illustrated in Figure 3.5(a), the highest stress concentrations are observed near the upper surface of the wheel hub, while the lowest stress occurs around the tie

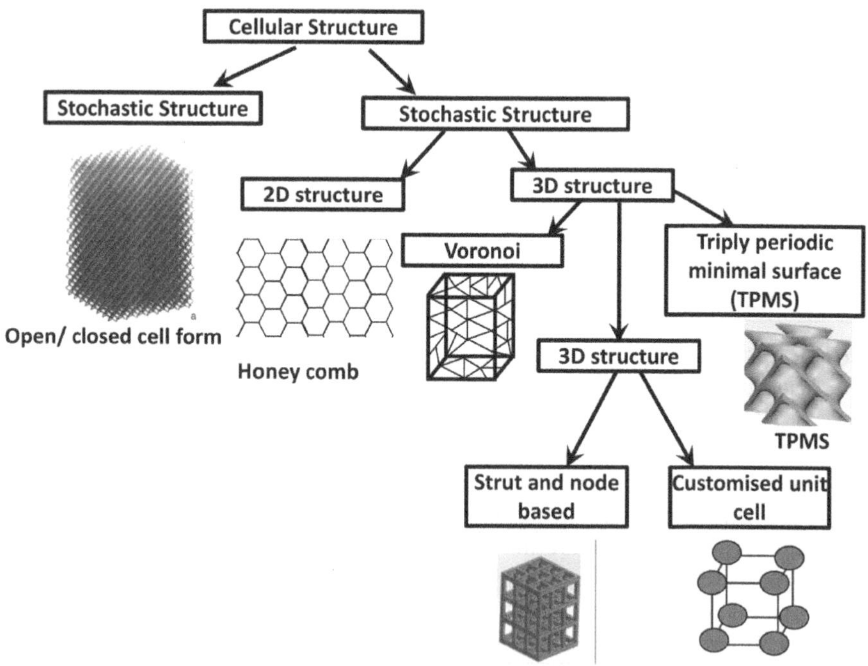

FIGURE 3.4 Various cellular structures and their configurations [4].

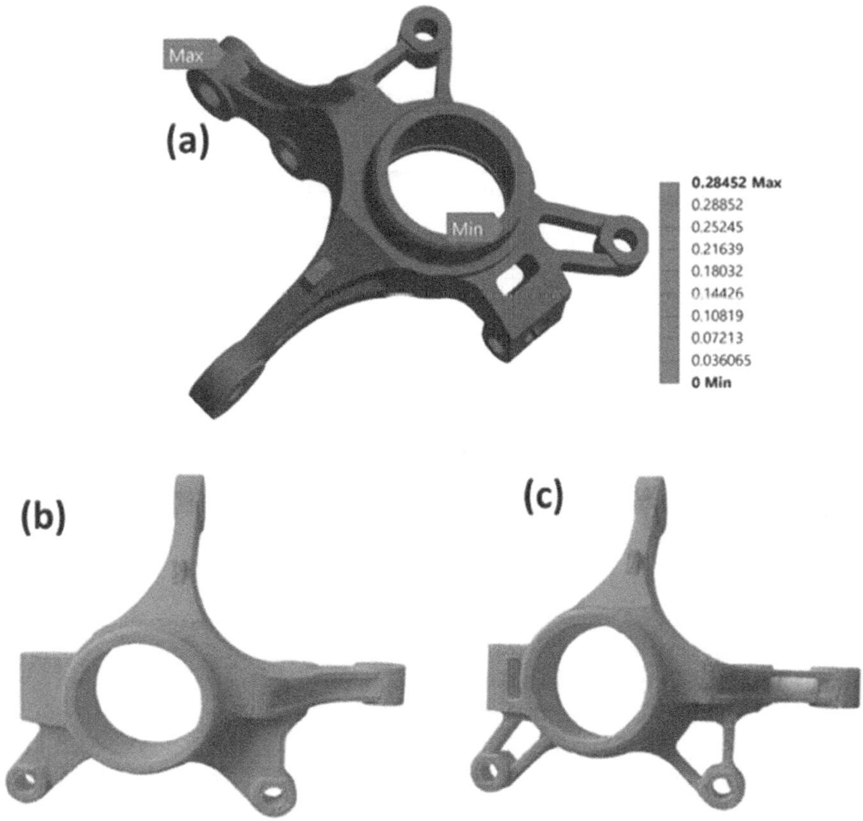

FIGURE 3.5 (a) Topology-optimized steering knuckle deformation stress; additive manufactured model of steering knuckle using FDM printer: (b) original and (c) topology-optimized versions [10].

rod mounting point. Through topology optimization aimed at reducing compliance, approximately 14.8% of material was removed from low-stress regions of the steering knuckle. Figure 3.5(b) and (c) depicts the steering knuckle fabricated via fused deposition modeling (FDM) printing, showing the original model in (b) and the topology-optimized version in (c).

3.3.2.1 Principles of Topology Optimization

Topology optimization is a mathematical methodology aimed at identifying the most efficient material layout within a defined design space to achieve desired performance objectives, typically reducing weight while enhancing stiffness or structural strength [9].

The typical workflow includes the following key steps:

- **Design Space Definition**: The allowable region for material placement is established, based on the part's intended function and physical constraints.

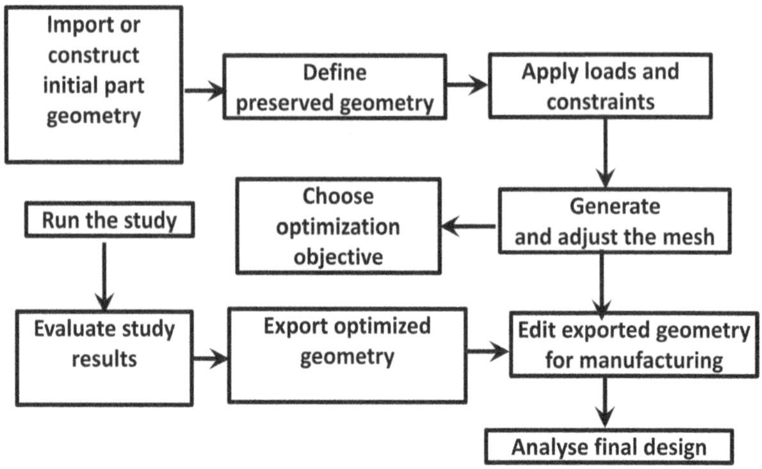

FIGURE 3.6 Steps for topology optimization [11].

- **Application of Loads and Boundary Conditions**: Realistic loading scenarios and constraints are applied to emulate actual working environments.
- **Objective Function and Constraints**: An objective function (such as minimizing compliance or mass) is specified, along with performance constraints (e.g., limits on stress or deflection).
- **Iterative Optimization Process**: Algorithms like the Solid Isotropic Material with Penalization or level-set methods are employed to iteratively adjust the material layout until an optimal solution is achieved.

The resulting structures often exhibit intricate geometries, such as internal lattices or cellular forms, that are difficult to manufacture using conventional techniques but are ideally suited for realization through AM technologies. Figure 3.6 shows the steps for the topology optimization [11].

3.3.3 Ribbed and Corrugated Structures

Thin-walled components reinforced with features such as ribs and corrugations exhibit significantly improved bending and torsional stiffness while retaining their lightweight characteristics. Incorporating these reinforcements into thin-walled designs offers multiple advantages, including enhanced load-bearing capacity, superior impact resistance, and overall mechanical performance.

Moreover, sandwich panels featuring lattice or honeycomb cores deliver exceptional impact resistance, making them especially suitable for demanding applications in aerospace and automotive industries. Both ribbed and corrugated structures provide unique benefits, making them suitable for different applications depending on the desired mechanical properties. In some cases, hybrid designs incorporating both features can be employed to achieve optimal performance.

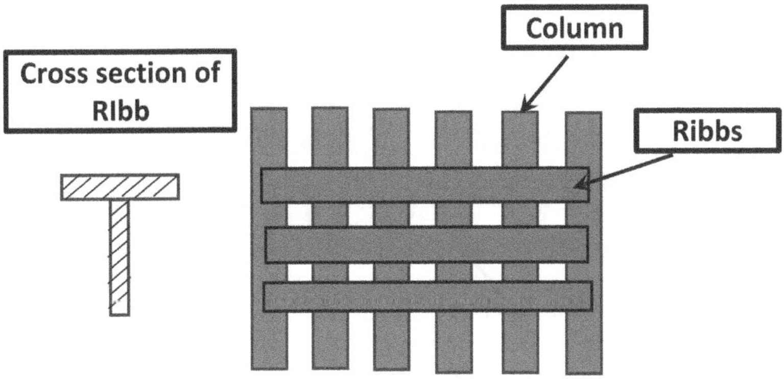

FIGURE 3.7 Schematic diagram of the ribbed structure [12].

3.3.3.1 Ribbed Structures in Additive Manufacturing

Ribbed structures involve the integration of strategically positioned reinforcements within thin-walled geometries to improve mechanical behavior. In the context of AM, ribbed designs offer the following benefits [12]:

- **Enhanced Stiffness and Strength:** The addition of ribs significantly boosts the structural integrity of components, minimizing deformation under applied loads.
- **Optimized Material Usage:** Ribs can be placed selectively in high-stress regions, improving strength without a substantial increase in overall material usage.
- **Improved Load Distribution:** These reinforcements promote uniform stress distribution, reducing localized stress concentrations and increasing fatigue life.
- **Design and Manufacturing Flexibility:** Advanced AM processes such as selective laser sintering (SLS) and fused deposition modeling (FDM) enable the fabrication of intricate ribbed geometries with high precision.

Ribbed structures are particularly advantageous in industries where weight reduction is critical without compromising structural integrity, such as the aerospace and automotive sectors. Figure 3.7 shows the schematic of the ribbed structure. Additionally, in biomedical engineering, ribbed configurations can enhance the mechanical reliability of implants and prosthetics while preserving a lightweight form factor.

3.3.4.2 Corrugated Structures in Additive Manufacturing

Corrugated structures are widely utilized in aerospace applications, such as fuselage panels and wing components, where structural efficiency and impact resistance are crucial. Corrugated structures incorporate wave-like patterns that enhance the mechanical performance of thin-walled components. Figure 3.8 shows the schematic of the corrugated structures. In the automotive industry, corrugated panels enhance crashworthiness and reduce the overall weight of vehicles [7].

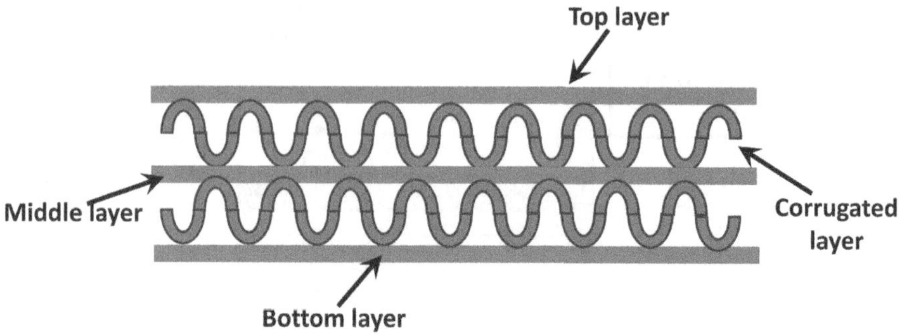

FIGURE 3.8 Schematic diagram of the corrugated structure [3,13].

The advantages of corrugated structures in AM include:

- **Enhanced Bending and Torsional Stiffness:** The corrugation pattern significantly improves resistance to bending and twisting forces.
- **High Energy Absorption:** Due to their unique deformation mechanisms, corrugated structures exhibit superior energy absorption capabilities, making them ideal for impact-resistant applications.
- **Optimized Weight-to-Strength Ratio:** Corrugations provide strength reinforcement without the need for additional material, leading to lightweight yet robust designs.
- **Versatile Geometries:** AM technologies enable the fabrication of complex corrugated patterns that are difficult to achieve through conventional manufacturing methods.

3.3.4 Bio-Inspired Structural Designs

Nature-inspired geometries—such as honeycomb, bone-mimetic, spider silk, and leaf venation patterns—enhance structural performance by emulating load-bearing systems that have been optimized through evolution. These biologically derived designs offer superior mechanical efficiency while minimizing material consumption, making them ideal for high-performance, lightweight applications [3].

- **Honeycomb Architectures**: Commonly found in beehives and insect shells, honeycomb patterns deliver an exceptional strength-to-weight ratio and high energy absorption capacity. These attributes make them well suited for aerospace and automotive components where both impact resistance and weight reduction are critical. Figure 3.9 shows the honeycomb structure of the walls for strength-to-weight ratio applications.
- **Bone-Mimetic Structures**: Modeled after the porous trabecular networks in human bone, these designs promote osseointegration and effective load distribution in biomedical implants. Their interconnected pore geometry not only facilitates bone ingrowth but also maintains a lightweight profile.

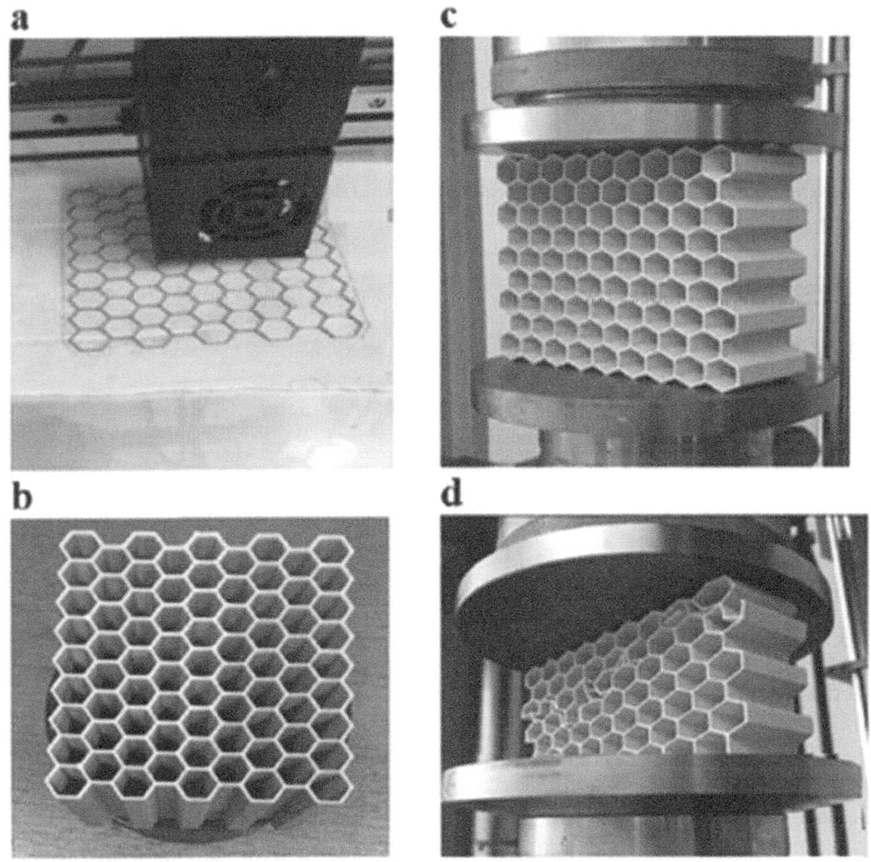

FIGURE 3.9 Honeycomb structure design of material architecture for additive manufacturing [9].

- **Spider Silk-Inspired Lattices**: Drawing inspiration from the remarkable tensile strength and elasticity of spider silk, these lattices provide superior toughness and energy dissipation properties, making them ideal for protective and impact-resistant applications.
- **Leaf Venation Networks**: Replicating the natural flow channels observed in leaf structures, these patterns efficiently distribute stress and optimize material usage, leading to lighter yet robust designs.

The adoption of bio-inspired designs in AM enables the fabrication of highly complex, multifunctional geometries tailored to specific mechanical requirements. The ability to fabricate these complex geometries through AM enables innovative solutions across various industries, including aerospace, biomedical, and structural engineering.

Each of these design strategies offers unique benefits that cater to specific industry requirements. Table 3.2 shows the comparison of ribbed, corrugated, and bio-inspired elements that can yield optimal performance [2].

TABLE 3.2
Comparative analysis of ribbed, corrugated, and bio-inspired structures [2]

Feature	Ribbed structures	Corrugated structures	Bio-inspired structures
Stiffness Improvement	High	Moderate	Variable (depends on design)
Energy Absorption	Moderate	High	Very High
Manufacturing Complexity	Moderate	High	High
Weight Optimization	High	Very High	Extremely High
Application Suitability	Aerospace, Automotive, Biomedical	Aerospace, Automotive	Aerospace, Biomedical, Structural

3.3.5 Functionally Graded Thin-Walled Structures

Functionally graded structures (FGS) are advanced engineering solutions characterized by a gradual variation in material composition and structural properties across a component. This spatially controlled transition enhances the mechanical performance of parts by aligning material distribution with varying load conditions, effectively reducing stress concentrations and improving structural efficiency.

3.3.5.1 Key Advantages of FGS in Additive Manufacturing

- **Tailored Mechanical Performance**: The smooth gradient in material properties enhances strength, toughness, and thermal resistance where needed, providing site-specific functionality within a single part.
- **Enhanced Stress Management**: Functionally graded materials minimize abrupt material interfaces, which helps in mitigating localized stress concentrations and improving fatigue life.
- **Lightweight Design with Superior Strength**: Strategic material distribution allows FGS to maintain high structural integrity while reducing overall mass—ideal for applications where weight savings are crucial.
- **Custom Thermal Behavior**: FGS can be engineered to gradually transition between thermally insulating and conductive materials, making them particularly valuable in aerospace and high-temperature environments.

3.3.5.2 Additive Manufacturing Techniques for Producing FGS

- **Gradient Material Deposition**: Processes like directed energy deposition (DED) and multi-material extrusion enable precise control over the deposition of dissimilar materials, facilitating smooth gradients.
- **Laser-Based Powder Bed Fusion (PBF)**: Technologies such as SLM and electron beam melting (EBM) allow for localized control over powder composition and melting parameters, supporting the fabrication of compositional gradients.

- **Gradient Density Design via Topology Optimization**: By integrating computational modeling and optimization algorithms, material density and mechanical properties can be spatially tailored to meet targeted performance goals.

FGS offer immense potential in pushing the boundaries of performance in critical sectors like aerospace, biomedical, automotive, and thermal management systems, where both precision and performance are non-negotiable.

3.4 STRATEGIES FOR MANUFACTURING THIN-WALLED STRUCTURES USING ADDITIVE MANUFACTURING

Manufacturing thin-walled structures through AM requires a combination of material selection, process optimization, and post-processing techniques to overcome challenges such as warping, instability, and surface defects [14]. Figure 3.10 shows the schematic of various strategies for the manufacturing development of thin-walled structures.

3.4.1 Material Selection

The choice of material is critical in AM, particularly for thin-walled structures. High-strength, low-thermal-expansion materials such as titanium alloys, aluminum alloys, and high-performance thermoplastics are commonly used due to their excellent mechanical properties and stability under thermal stress. These materials help minimize warping and dimensional instability, which are common challenges in thin-walled fabrication. Additionally, it is important to ensure uniform properties across the material, such as consistent particle size or filament quality powder. This consistency is crucial for achieving reliable, repeatable results in the AM process and ensuring the structural integrity of the final product.

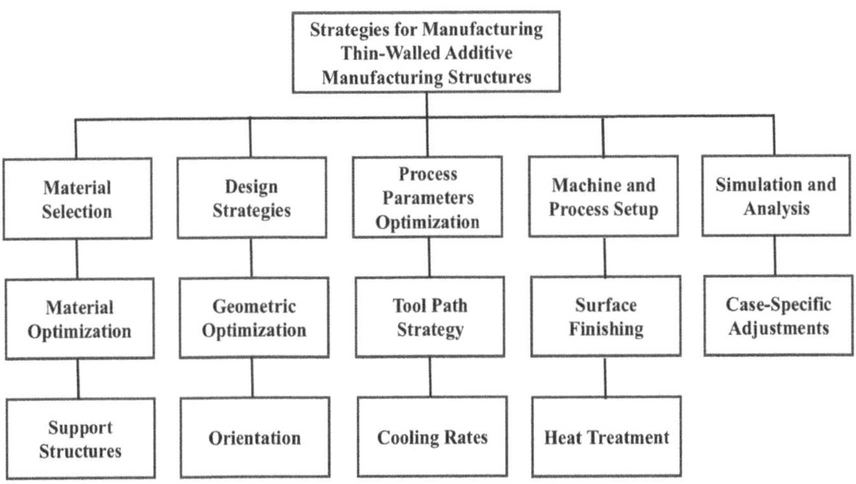

FIGURE 3.10 Schematic of various strategies for manufacturing development of thin wall structures [9].

3.4.2 Design Strategies

Designing thin-walled structures in AM requires careful planning to ensure stability, strength, and manufacturability. The following strategies can help optimize the design for successful fabrication:

- **Wall Thickness:** Identify the minimum feasible wall thickness that the AM process can reliably produce. This depends on factors such as the laser spot size in PBF or nozzle diameter in extrusion processes, as well as the material properties being used. Maintaining a consistent and adequate thickness ensures structural integrity and minimizes the risk of failure during or after fabrication.
- **Geometric Optimization:** Employ techniques like topology optimization and lattice structures to reduce weight without compromising strength. These methods enable the redistribution of material only where it is needed, creating efficient, lightweight designs that are ideal for thin-walled applications.
- **Support Structures:** Optimize or minimize the use of support structures to reduce material consumption, printing time, and post-processing effort. For example, strategic placement of supports in necessary regions and eliminating unnecessary ones can enhance both manufacturability and sustainability.
- **Orientation:** Position the part on the build platform to minimize overhangs and improve the resolution of thin-walled areas. Proper orientation reduces the reliance on supports, prevents distortions, and ensures better surface finishes, particularly in areas requiring high precision.

3.4.3 Process Parameters Optimization

Optimizing process parameters is crucial for achieving high-quality thin-walled structures in AM. Below are key strategies for fine-tuning the process:

- **Layer Thickness:** Utilize thinner layers to enhance surface finish and dimensional accuracy. Reduced layer thickness not only provides finer detail but also helps in maintaining consistency across thin-walled sections, particularly in intricate geometries.
- **Energy Input:** Precisely control laser power, scan speed, or extrusion rate based on the material and AM technique. Proper energy input ensures uniform melting or deposition, preventing defects such as overheating, overbuild, or material underutilization that can compromise thin-wall stability.
- **Tool Path Strategy:** Adopt specialized tool path strategies like fine contour scanning or spiral paths to minimize edge and corner defects. These strategies ensure a smoother deposition and reduce irregularities, particularly in areas with sharp transitions or delicate features.

- **Cooling Rates:** Optimize cooling rates to mitigate thermal stresses and prevent warping or cracking, which are especially critical in metal-based AM. Controlled cooling improves the structural integrity and dimensional stability of thin-walled parts.

3.4.4 Machine and Process Setup

Setting up the machine and process appropriately is essential for the successful fabrication of thin-walled structures in AM. Key strategies include:

- **Precision Equipment:** Utilize high-resolution AM machines designed for fine-detail fabrication. Such equipment ensures minimal deviation and is capable of producing intricate thin-walled features with high dimensional accuracy.
- **Nozzle and Beam Diameter:** Select nozzles and laser or beam diameters that are optimized for thin-wall production. Smaller diameters enable finer deposition or scanning, resulting in improved resolution and better control over the wall thickness.
- **In-Situ Monitoring:** Integrate real-time monitoring systems to inspect each layer during the build process. These systems detect and address deviations, defects, or anomalies early, ensuring consistent quality throughout the fabrication of thin-walled structures.

3.4.5 Post-Processing Techniques

Effective post-processing is crucial for enhancing the surface quality, mechanical properties, and dimensional accuracy of thin-walled structures in AM. Key strategies include:

- **Surface Finishing:** To improve surface quality and dimensional accuracy, employ techniques like polishing, chemical etching, or abrasive blasting. These methods smooth the surface, remove roughness, and enhance the aesthetic and functional properties of the thin walls, making them more suitable for high-performance applications.
- **Heat Treatment:** Use controlled heat treatment processes such as annealing or aging to relieve residual stresses and enhance the mechanical properties of the material. These treatments improve material strength, hardness, and fatigue resistance, ensuring better performance and durability in the final product.
- **Dimensional Tuning:** For parts requiring high precision, use Computer Numerical Control (CNC) machining or grinding to fine-tune critical dimensions. These subtractive processes allow for precise adjustments to achieve the required tolerances, ensuring that thin-walled structures meet stringent specifications.

3.4.6 Simulation and Analysis

Effective simulation and analysis are essential for ensuring the success and reliability of thin-walled structures in AM. Key strategies include:

- ***Finite Element Analysis (FEA):** Use FEA to model stress distributions and thermal gradients within the structure. This helps predict potential deformations, weaknesses, or failure points, enabling preemptive design adjustments for enhanced structural integrity.
- **Process Simulation:** Conduct simulations of the AM process to evaluate and optimize parameters specifically for thin-walled features. These simulations provide insights into layer deposition behavior, heat transfer, and material interactions, ensuring better process control and reduced risk of defects.
- **Quality Control:** Implement non-destructive testing techniques such as X-ray computed tomography or ultrasonic inspection to detect internal and external defects. These methods ensure the integrity of thin-walled structures without compromising their usability.

3.4.7 Case-Specific Adjustments

For optimized production of thin-walled structures, case-specific adjustments can significantly enhance performance, precision, and material efficiency. Key strategies include:

- **Hybrid Manufacturing:** Integrate AM with subtractive methods (such as CNC machining) for high-precision thin walls. This hybrid approach allows for the creation of complex geometries through AM while ensuring that critical areas are refined to meet tight tolerances, improving both surface finish and dimensional accuracy.
- **Multi-Material Printing:** Utilize multi-material or gradient material printing to combine different materials in a single structure. This technique can optimize material properties, offering improved strength in specific areas while minimizing material usage. For example, combining a strong material on the exterior with a lighter or more flexible material on the interior can achieve a balance of performance and weight reduction.

3.5 PREFERRED ADDITIVE MANUFACTURING PROCESSES

The selection of AM processes depends on material and application requirements. Selective laser powder bed fusion (LPBF) and EBM are ideal for high-resolution metal lattices, while WAAM is suitable for large-scale, ribbed, or graded structures with controlled bead overlapping. Table 3.3 shows the comparison of the AM process based on its performance matrix. By integrating these design strategies, engineers can fabricate highly efficient, lightweight, thin-walled structures optimized for mechanical strength, durability, and manufacturability.

TABLE 3.3
Comparison of additive manufacturing processes based on key performance metrics [14]

AM process	Dimensional accuracy	Surface finish	Mechanical properties	Build rate	Cost efficiency
WAAM	Moderate	Poor	High	Very High	High
SLM	High	Good	Excellent	Moderate	High
DMLS	High	Good	Excellent	Moderate	High
EBM	Moderate	Moderate	High	High	Moderate
FDM	Low	Poor	Moderate	High	Very High
SLA	Very High	Excellent	Low to Moderate	Low	Moderate

TABLE 3.4
Comparison of various AM techniques for thin-walled polymers printing [16–17]

Process	Min. wall thickness	Resolution	Material types	Applications
FDM	~0.4 mm	Moderate	PLA, ABS, PETG, ULTEM	Prototyping, tooling, enclosures
SLA	~0.1 mm	High	Photopolymer resins	Microfluidics, dental, biomedical
DLP	~0.1 mm	High	Photopolymer resins	Precision prototypes, scaffolds
SLS	~0.5 mm	Moderate	Nylon, TPU	Functional parts, aerospace, tooling
Material Jetting	~0.1 mm	Very High	Multi-material resins	Biomedical models, complex geometries

3.6 ADDITIVE MANUFACTURING PROCESSES FOR THIN-WALLED STRUCTURES

3.6.1 POLYMER STRUCTURES

Thin-walled polymer components, defined by their reduced wall thickness and high aspect ratio, are increasingly being produced through AM technologies. The inherent advantages of AM, such as design flexibility, high precision, and layer-by-layer construction, make it particularly well suited for creating complex geometries that are difficult to achieve with conventional manufacturing methods. Table 3.4 provides a comparative overview of different AM techniques used for fabricating these structures. Figure 3.11 demonstrates a real-world application in the shipbuilding industry, where AM enables the production of lightweight, robust, and geometrically intricate ship hull components.

FIGURE 3.11 Schematic diagram illustrating a thin-walled polymer-structured ship hull, showcasing the application of additive manufacturing in shipbuilding [15].

3.6.2 Metal Structures

Thin-walled metallic components are critical to industries such as aerospace, automotive, energy, and biomedical engineering, and they require high precision, controlled microstructure, and exceptional mechanical integrity. AM technologies have emerged as transformative tools for fabricating these components, offering unparalleled advantages in design freedom, material efficiency, and the ability to produce complex geometries. Table 3.5 provides a comparative analysis of various AM techniques employed in the fabrication of thin-walled metallic structures. Figure 3.12(a) illustrates a monolithic Ti-6Al-4V tibial (knee) stem prototype developed for biomedical applications [18].

Figure 3.12(b) showcases a thin-walled component fabricated from 316L stainless steel, highlighting the potential of AM in enabling lightweight, durable, and geometrically complex metallic structures [19].

3.6.3 Ceramic Structures

Ceramic materials are renowned for their outstanding properties, including high hardness, thermal stability, corrosion resistance, and biocompatibility. Despite these advantages, their inherent brittleness and machining challenges pose significant limitations for conventional manufacturing methods. AM has emerged as a promising solution for overcoming these constraints, enabling the fabrication of thin-walled and complex ceramic components with high precision. This makes AM particularly valuable in high-performance applications across sectors such as aerospace, biomedical,

TABLE 3.5
Additive manufacturing techniques for thin-walled polymers [16,17]

Process	Min. wall thickness	Resolution	Key materials	Applications
LPBF / SLM	~150–200 μm	High	SS, Ti, Al, Inconel, CoCr	Aerospace, biomedical implants, microstructures
EBM	~300–500 μm	Moderate	Ti-6Al-4V, Inconel, CoCr	Orthopedic implants, structural aerospace parts
DED / WAAM	>1 mm	Low–Moderate	SS, Ti, Al, Ni alloys	Large structures, pressure vessels, tooling
Binder Jetting	>0.5 mm (after sintering)	Moderate	SS, bronze, tool steel	Industrial components, tooling, prototypes
Cold Spray AM	~0.5–1 mm	Low	Al, Cu, Ti	Repair, coatings, lightweight reinforcement

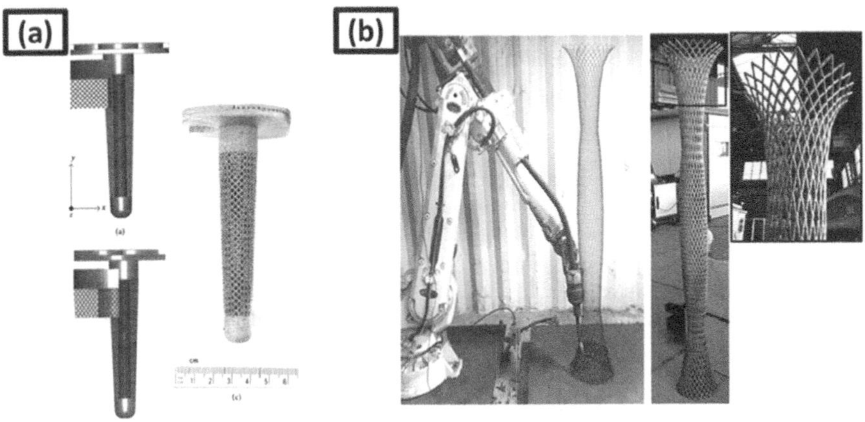

FIGURE 3.12 (a) Monolithic Ti-6Al-4V tibial (knee) stem prototype developed using additive manufacturing for biomedical applications [18] and (b) thin-walled outer shell structure fabricated from 316L stainless steel, demonstrating the potential of AM in producing lightweight, durable, and geometrically complex components [19].

electronics, and energy. Table 3.6 presents a comparative analysis of various AM techniques used in the fabrication of thin-walled ceramic structures. Figure 3.13 illustrates additively manufactured yttria-stabilized tetragonal zirconia polycrystal dental bridges produced via ceramic stereolithography, along with their corresponding microstructures, which reveal features such as pores and grain growth [20].

TABLE 3.6
Comparative analysis of various AM techniques for fabrication of thin-walled ceramic structures

Process	Min. wall thickness	Resolution	Common materials	Applications
SLA / DLP	~50–200 μm	High	Zirconia, Alumina, Hydroxyapatite	Dental crowns, microfluidics, implants
Robocasting / DIW	~300–500 μm	Moderate	Alumina, SiC, Bioceramics	Filters, bone scaffolds, catalyst supports
Binder Jetting	~500–1000 μm	Moderate	Alumina, SiC, Zirconia	Housings, porous ceramics, high-temp components
SLS / LIS	~200–500 μm	Moderate	Silicon Nitride, Alumina	Refractories, engine parts, wear-resistant coatings
Inkjet Printing	~100–300 μm	High	PZT, Alumina, Ferrites	Sensors, micro-devices, electronics

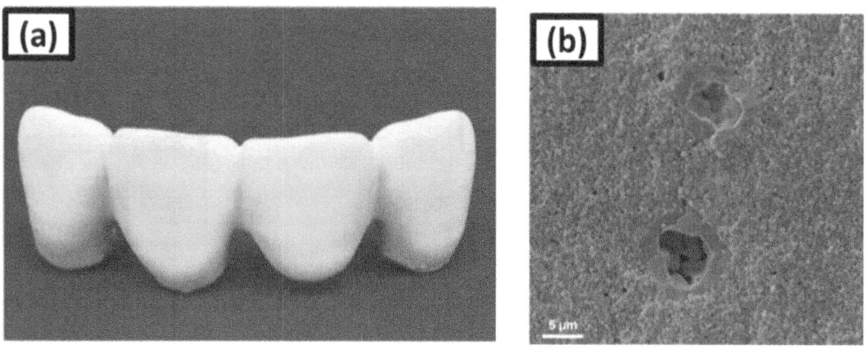

FIGURE 3.13 (a) Additively manufactured yttria-stabilized tetragonal zirconia polycrystal (Y-TZP) dental bridges by ceramic stereolithography and (b) Y-TZP products with pores and grown grains [20].

In summary, lightweight thin-walled structures play a pivotal role in industries such as aerospace and automotive, where reducing weight without compromising mechanical strength is a fundamental design objective. Critical components like airframes, fuel tanks, heat exchangers, and turbine blades rely on such structures to enhance performance, fuel efficiency, and overall system reliability. In the biomedical field, thin-walled designs are essential for developing custom prosthetics and implants, providing the necessary combination of mechanical strength, reduced weight, and enhanced patient comfort and long-term durability.

Despite their advantages, consistently fabricating high-quality thin-walled structures remains a significant challenge. Common issues include heat accumulation, residual stresses, warping, cracking, layer delamination, and microstructural inconsistencies, all of which can detrimentally impact structural integrity and part performance.

Each material class presents unique challenges, particularly concerning thermal behavior, interlayer bonding, and post-processing needs. For instance, metals may face severe residual stress and distortion due to thermal gradients, while polymers often struggle with layer adhesion, and ceramics are prone to cracking due to their brittle nature.

To mitigate these challenges, several strategies have been developed, including:

- Process parameter optimization to fine-tune energy input and deposition rates
- Thermal management techniques to reduce residual stresses and control cooling rates
- In-situ monitoring systems to detect and correct anomalies during the build process

Various AM techniques are employed in fabricating thin-walled structures, such as:

- WAAM
- SLM
- DMLS
- EBM

These methods are evaluated based on their suitability for producing thin-walled components, considering factors like material compatibility, geometric complexity, mechanical performance requirements, and post-processing needs.

By thoroughly understanding and addressing these variables, manufacturers can reliably produce thin-walled structures that meet the stringent demands of aerospace, automotive, and biomedical applications.

3.7 CONCLUSION

Thin-walled structures play a vital role in applications demanding both lightweight design and high structural integrity, particularly in aerospace, automotive, and biomedical sectors. AM enables the fabrication of such complex geometries with enhanced design flexibility and material efficiency. This chapter explored the challenges and advancements in producing thin-walled components using AM techniques across metals, polymers, and ceramics. Each material class poses distinct challenges—such as thermal stress, delamination, and dimensional inaccuracy—necessitating tailored strategies. Advanced AM processes like WAAM, SLM, DMLS, and EBM, along with optimized parameters and in-situ monitoring, provide effective solutions. Continued innovation in AM technology and material-specific processing will be essential for the reliable and scalable production of high-performance thin-walled structures in critical engineering applications.

Acknowledgments: One of the authors, N.H., thanks Mr. Akash Vincent for his support in collecting relevant information.

REFERENCES

1. A. Seharing, A.H. Azman, S. Abdullah, A review on integration of lightweight gradient lattice structures in additive manufacturing parts, *Adv Mech Eng* 12 (2020) 1–21.
2. W.E. Frazier, Metal additive manufacturing: A review, *J Mater Eng Perform* 23 (2014) 1917–1928.
3. C.W. Isaac, F. Duddeck, Current trends in additively manufactured (3D printed) energy absorbing structures for crashworthiness application – A review, *Virtual Phys Prototyp* 17 (2022) 1058–1101.
4. M. Iwaniszyn, Periodic open cellular structures (POCS) as catalyst supports—A review, *Energies (Basel)* 15 (2022) 1–18.
5. D. Moreno Nieto, D. Moreno Sánchez, M. Mandolini, Design for additive manufacturing: Tool review and a case study, *Appl Sci* 11 (2021) 1–12.
6. M. Francis, N. Kemper, Y. Maghdouri-White, N. Thayer, Additive Manufacturing for Advanced Applications, in: J. Zhang, Y.-G. Jung (Eds.), *Additive Manufacturing Materials, Processes, Quantifications and Applications*, Butterworth-Heinemann, 2018: pp. 311–344.
7. E. Uribe-Lam, C.D. Treviño-Quintanilla, E. Cuan-Urquizo, O. Olvera-Silva, Use of additive manufacturing for the fabrication of cellular and lattice materials: A review, *Mater Manuf Process* 36 (2021) 257–280.
8. X. Zhang, Z. Ni, A. Takezawa, Y. Wang, Z. Kang, Microscopic stress-constrained two-scale topology optimisation for additive manufacturing, *Virtual Phys Prototyp* 20 (2025) 1–26.
9. O. Ibhadode, Z. Zhang, J. Sixt, K.M. Nsiempba, J. Orakwe, A. Martinez-Marchese, O. Ero, S.I. Shahabad, A. Bonakdar, E. Toyserkani, Topology optimization for metal additive manufacturing: Current trends, challenges, and future outlook, *Virtual Phys Prototyp* 18 (2023):1–53.
10. S. Akilan, S. Janardhan Reddy, Design improvement of steering knuckle through topology optimization for additive manufacturing, *Cogent Eng* 11 (2024) 1–10.
11. D. Vlah, R. Žavbi, N. Vukašinović, Evaluation of topology optimization and generative design tools as support for conceptual design, in: *Proceedings of the Design Society: DESIGN Conference*, Cambridge University Press, 2020: pp. 451–460.
12. Q. Wang, C. Qian, Z. Wu, Research on the rational design method of strength reinforcement for thin-walled structure based on limit load analysis, *Appl Sci (Switzerland)* 12 (2022) 1–11.
13. H. Tetsuka, S.R. Shin, Materials and technical innovations in 3D printing in biomedical applications, *J Mater Chem B* 8 (2020) 2930–2950.
14. X. Peng, L. Kong, J.Y.H. Fuh, H. Wang, A review of post-processing technologies in additive manufacturing, *J Manuf Mater Process* 5 (2021) 1–23.
15. G.D. Goh, K.K. Wong, N. Tan, H.L. Seet, M.L.S. Nai, Large-format additive manufacturing of polymers: A review of fabrication processes, materials, and design, *Virtual Phys Prototyp* 19 (2024) 1–34.
16. G. Poologasundarampillai, A. Nommeots-Nomm, Materials for 3D printing in medicine: Metals, polymers, ceramics, hydrogels, in: *3D Print Med*, Elsevier Ltd, 2017: pp. 43–71.
17. S.C. Ligon, R. Liska, J. Stampfl, M. Gurr, R. Mülhaupt, Polymers for 3D printing and customized additive manufacturing, *Chem Rev* 117 (2017) 10212–10290.
18. L.E. Murr, S.M. Gaytan, E. Martinez, F. Medina, R.B. Wicker, Next generation orthopaedic implants by additive manufacturing using electron beam melting, *Int J Biomater* 4 (2012) 1–14.

19. S.C.A. Costello, C.R. Cunningham, F. Xu, A. Shokrani, V. Dhokia, S.T. Newman, The state-of-the-art of wire arc directed energy deposition (WA-DED) as an additive manufacturing process for large metallic component manufacture, *Int J Comput Integr Manuf* 36 (2023) 469–510.
20. H. Li, L. Song, J. Sun, J. Ma, Z. Shen, Dental ceramic prostheses by stereolithography-based additive manufacturing: Potentials and challenges, *Adv Appl Ceram* 118 (2019) 30–36.

4 Modeling and Simulation of Additively Manufactured Thin-Walled Structures

Sagar Nikam, Neelesh Kumar Jain, and Mayur Sawant

4.1 INTRODUCTION

Additive manufacturing (AM) is a bottom-up manufacturing approach possessing ability to (i) manufacture complex geometries; (ii) add delicate features to an existing product; (iii) remanufacture, repair, or refurbish a defective and/or worn but usable product; and (iv) customize a product. Its evolution started at the beginning of the 21st century and is still continuing. It is referred to by different names, such as AM, layered manufacturing, additive fabrication, and solid freeform fabrication [1]. Most of the AM processes manufacture near-net-shape components directly from data of their computer-aided design models with minimum loss of material. Therefore, AM has evolved as a material-efficient, energy-efficient, and environmentally friendly alternative to the top-down manufacturing philosophies such as subtractive manufacturing (i.e., material removal processes), in which significant material is wasted, and deformative manufacturing (i.e., rolling, forging, extrusion, drawing, deep drawing, sheet-metal forming, etc.), which is energy-intensive. Furthermore, it eliminates many steps required in conventional manufacturing by utilizing the knowledge of product modeling, robotics, joining, sensing, monitoring, and controlling. Initially, use of the AM processes was limited to low-melting-point materials such as polymers and polymer matrix composites. With sustained research and industry focus, applications of AM have been extended to produce functionally graded materials [2], porous structures [3], and low-melting-point metallic materials. AM processes are being continuously used for different applications in the fields of aerospace, automotive, biomedical, nuclear, electronics, tool-making, special-purpose engineering, and repair or remanufacturing industries.

DOI: 10.1201/9781003571209-4

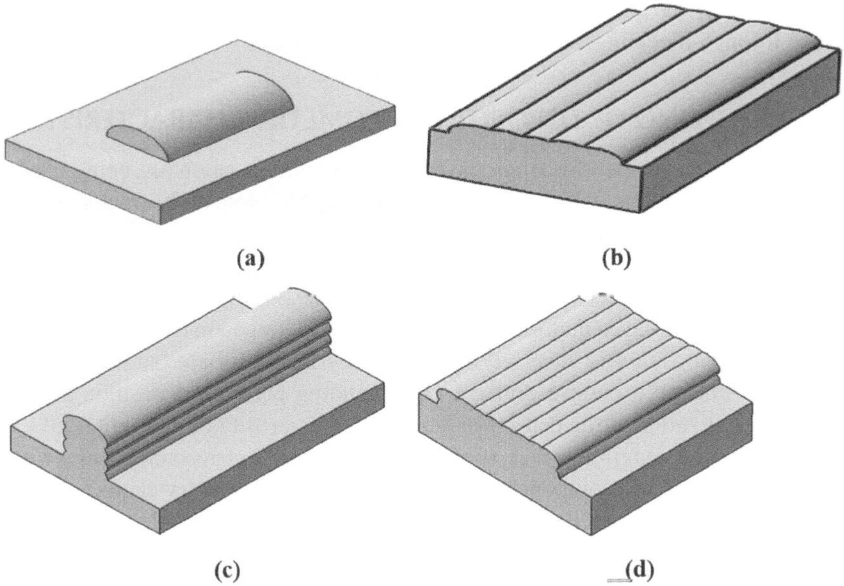

FIGURE 4.1 Concept of layer and track deposition used in an AM process: (a) single-layer single-track deposition, (b) single-layer multi-track deposition, (c) multi-layer single-track deposition, and (d) multi-layer multi-track deposition [4]. AM, Additive manufacturing.

Figure 4.1 presents the concept of layer and track deposition used in an AM process. Their following different combinations are used for different applications of an AM process: (i) single-layer single-track (Figure 4.1(a)) deposition, which is typically used for repairing and remanufacturing applications; (ii) single-layer multi-track (Figure 4.1(b)) deposition, in which tracks are deposited with some overlap to cover the entire top surface of the substrate material. It is typically used for various coating applications; (iii) multi-layer single-track (Figure 4.1(c)) deposition in which layers are deposited successively to achieve the desired height. It is typically used in the manufacturing of thin-walled complex geometries; and (iv) multi-layer multi-track (Figure 4.1(d)) deposition in which layers are deposited along the width and height of the substrate to achieve the desired width and height of complex geometries typically used in cladding applications. For multi-layer depositions (Figure 4.1(c) and (d)), the previously deposited layer becomes the substrate for deposition of the next layer, except for the deposition of the very first layer.

However, the applicability of AM processes to high-melting-point metallic materials, ceramics, and biocompatible materials is still challenging and has been attracting the interest of the manufacturing research community and manufacturing industries. Researchers at various renowned universities and research laboratories are continuously engaged in (i) developing new AM processes that can economically manufacture complex-shaped components made of biocompatible materials (such as alloys of titanium, nickel, and cobalt), shape memory materials (such as Ni-Ti alloy), and high-melting-point metallic materials; (ii) enhancing material and shape applicability of the existing AM processes; and (iii) improving the process performance through modeling, simulation, and optimization.

4.2 MODELING OF DEPOSITION GEOMETRY CHARACTERISTICS

AM processes developed for single-layer single-track deposition are being used for (i) coating applications, which generally require single-layer multi-track deposition with overlap between successive tracks; (ii) freeform fabrication of a typical component, which generally requires multi-layer single-track deposition with very good metallurgical bonding between the successive layers; and (iii) cladding, which generally requires multi-layer multi-track deposition [5]. The geometry of a single-layer single-track deposition is adversely influenced by inappropriate parameters of the deposition process, excessive dilution, poor bonding between deposition and substrate, low utilization of the deposition material, and porosity of deposition [6].

The bonding strength between successive layers and dimensional accuracy of an additive-manufactured component are influenced by the geometry characteristics of the single-layer single-track deposition. These characteristics include shape of deposition (i.e., semi-circular, elliptical, or parabolic), deposition height 'h', deposition width 'w', aspect ratio (i.e., ratio of deposition width 'w' to deposition height 'h'), dilution (i.e., ratio of area of diluted area 'A_2' to summation of deposition area 'A_1' and diluted area 'A_2'), and deposition root angle 'θ' as depicted in Figure 4.2. For good quality deposition, the bead root angle should lie in the range from 60° to 80°, aspect ratio in the range from 1 to 4, and dilution in the range from 6% to 10% [7].

Three distinct zones exist in a substrate material, namely melt pool zone, solid–liquid interface, and the solid substrate zone. Formation of a melt pool takes place when an intense amount of heat is applied on the surface of a substrate material using a suitable heat source. Dimensions of a melt pool depend on the amount of heat conducted away from the solid–liquid interface of the melt pool to the solid substrate material, as illustrated in Figure 4.2. The following sections describe the different models developed to predict the melt pool dimensions and deposition geometry characteristics for different AM processes.

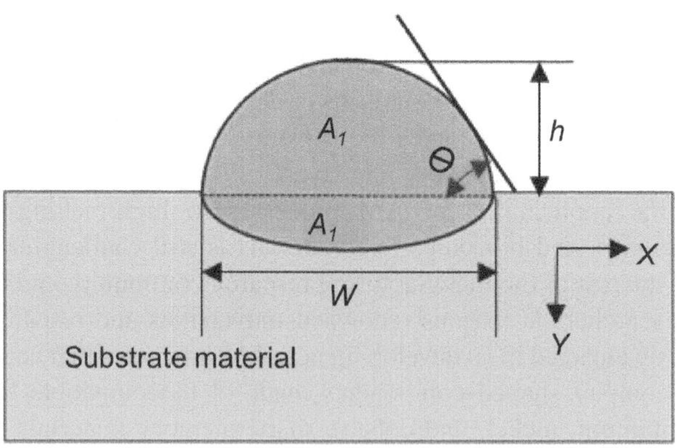

FIGURE 4.2 Geometry of a typical single-layer single-track deposition [8].

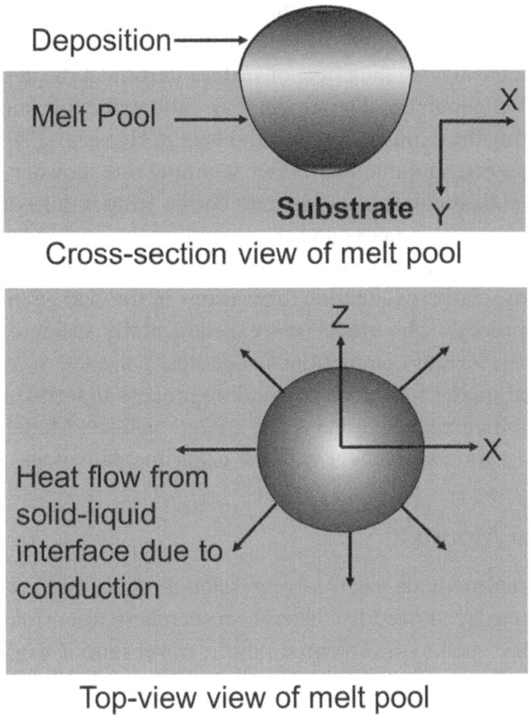

FIGURE 4.3 Formation of the melt pool in a substrate material.

4.2.1 Thermal Modeling

Following are brief details of the past work done on the development of thermal models for the prediction of deposition characteristics in different AM processes.

Bamberger et al. [9] developed a thermal model in terms of the process parameters of a laser-based deposition process to predict geometry characteristics of surface alloying of AISI 4130 steel using Ni_2B and CrB powders. They obtained good agreement between the model-predicted values and the experimental results. Picasso et al. [10] developed a mathematical model in terms of laser power, travel rate, and powder flow rate to determine the melt pool dimensions in the laser cladding process. They assumed an elliptical melt pool cross-section, with the melt pool depth equal to the clad height. In contrast, for the same process, Dobranich and Dykhuizen [11] modeled the melt pool as hemispherical, corresponding to a circular cross-section. Pinkerton and Li [6] used a steady-state condition to determine the dimensions of the melt pool in the laser cladding process by balancing the amount of heat conducted through the perimeter of the melt pool and considering its cross-section as circular. Their model did not consider powder flow rate for the prediction of the geometry characteristics. Liu and Li [12] developed a theoretical model as a function of laser scanning velocity, powder flow rate, and laser power to predict the width of a thin wall by the laser cladding process with an objective to determine the minimum laser power required to fabricate a thin wall of metallic materials. They validated their model experimentally for a thin wall through multi-layer single-track deposition. Lalas et al. [13] developed an analytical model as a function of torch travel speed and powder feed rate, considering the effect of surface tension of the deposition geometry to predict laser clad geometry. The model was validated experimentally and found capable of predicting the deposition width and height. Hua et al. [14] studied the influence of the laser power, spot diameter, laser scanning rate, powder feed rate, powder feeding position, and shielding gas flow rate on the temperature of the melt pool in the laser-based deposition process and developed a relationship between deposition thickness and melt pool temperature. Cheikh et al. [15] developed a mathematical model to predict the clad cross-section dimensions of the clad geometry in the direct laser fabrication process. The model was experimentally validated by depositing a number of tracks for various combinations of process parameters. Ya et al. [16] developed a 2D thermal model for the laser cladding process in terms of process parameters and powder efficiency using mass and energy balance. Model-predicted values were found to be in good agreement with the experimental values.

4.2.2 Empirical Modeling

Empirical modeling methods, such as regression analysis and artificial neural networks (ANNs), have been used by different researchers to correlate the deposition process parameters, such as heat source power, travel rate of worktable, and deposition material feed rate, with deposition characteristics such as deposition width, height, and dilution measured experimentally. Accuracy of the developed empirical models mainly depends on the number of experimental runs conducted. The following is a summary of past work done on empirical modeling for prediction of deposition characteristics in different AM processes.

Nagesh and Datta [17] used ANN to model the deposition geometry and dilution in terms of parameters of the shielded metal arc deposition process. Palani and Murugan [18] developed regression equations in terms of parameters of the fluxed core arc-based deposition process to predict the deposition geometry characteristics. They considered the individual and interaction influence of the process parameters while developing the regression model. Onwubolu et al. [19] developed a first-order regression equation considering only one input parameter and a second-order regression equation considering multiple combinations of input parameters to predict the cladding angle. Davim et al. [20] used multiple regression analysis to predict geometry characteristics of the deposition by the laser cladding process. They studied influence of laser power, feed rate of material, and travel speed of laser on the geometry of clad and found the prediction error to be in the acceptable range. Sun and Hao [21] developed empirical equations relating clad width, clad height, and dilution with laser cladding process parameters and their interactions during deposition of Ti-6Al-4V powder on a similar substrate material. Nenadl et al. [22] developed regression models to correlate clad width, height, and overlap between successive tracks with parameters of the laser cladding process. Model-predicted clad width, height, and overlap showed good agreement with their experimentally measured values. Xiong et al. [23] developed models using ANN and second-order regression analysis to study the influence of size, quality, and accuracy of single-track deposition by gas metal arc (GMA)-based AM. Saqib et al. [24] used ANN to model laser-based powder deposition of steel on a low-carbon substrate material. Lee et al. [25] formulated regression equations to predict deposition geometry and dilution depth as a function of parameters of the laser direct energy-based deposition process. They used the response surface method to design and conduct experiments and analyzed the effects of each process parameter on the cross-sectional profile of single-layer geometry. Model-predicted results were found to be in good agreement with measured responses. Nabhani et al. [26] used linear regression analysis to develop an empirical-statistical relation between characteristics of single-clad tracks (i.e., clad height, clad width, penetration depth, wetting angle, and dilution) and key parameters of the laser cladding process (i.e., power, travel speed, and deposition material feed rate). Experimental validation confirmed that the developed relations can predict the geometrical characteristics of the single-clad tracks with good accuracy.

4.2.3 Finite-Element Simulation of Melt Pool

Melt pool dimensions are significantly affected by the thermal properties of the substrate and deposition material, flow of molten material, type and amount of heat energy required for melting the deposition material, interaction duration of heat source with substrate, heat conducted within the substrate material, and heat losses to the surroundings. This makes the study of the influence of all these aspects on deposition characteristics dynamic and highly complicated and difficult to develop realistic theoretical models. Finite element simulation (FES) has been found to be a very effective tool in modeling and predicting deposition characteristics. Following is a brief description of the past work done on FES for prediction of deposition characteristics in different AM processes.

Hoadley et al. [27] developed a transient 3D-FE model to determine the flow of heat during the laser deposition process. Model-predicted results were obtained considering the dependency of melt pool dimensions on laser power and traverse speed and neglecting the deposition material addition to the melt pool. Picasso and Hoadley [10] used a two-dimensional finite element (2D-FE) model for the laser melting and cladding process. They found that the shape and size of the melt pool and the temperature distribution within the melt pool are mainly influenced by heat flow, Marangoni flow effects induced by surface tension gradients, and injection of deposition material in powder form. Kim and Peng [28] performed 2D-FES to determine the change in the melt pool shape by assuming that the width of the melt pool formed in the substrate material is equal to the width of the deposition track. For the same process, Lei et al. [29] developed a 2D-FE model to determine the change in melt pool shape caused by the Marangoni force and surface tension. Toyserkani et al. [30] did transient 3D-FES to determine deposition geometry characteristics for the laser cladding process. Melt pool size was determined from the temperature fields without the addition of deposition material. A similar FE model was developed by Han et al. [31] to simulate the addition of powder injection through a coaxial nozzle for the laser cladding process. The heat flow in the melt pool was simulated considering energy balances at the solid–liquid interfaces along with melting, solidification, and evaporation phenomena. Simulation was used to investigate the influence of powder injection and without powder injection on the melt pool shape.

4.3 THERMAL ANALYSIS OF THIN-WALLED STRUCTURES

Study of the distribution of temperature, temperature gradient, and thermal cycles during metallic deposition is essential in optimizing and selecting the required amount of heat input, which helps in the prediction of thermal distortion and residual stresses developed in the substrate and the deposited material. Optimizing the required heat input will result in minimum thermal distortion and thermal stresses, and the selection of optimum process parameters for different combinations of substrate and deposition materials will reduce extensive experimentation, which is very time-consuming, costly, and most of the time infeasible. Simulation of deposition in any AM process can solve these issues to a large extent. Studies have been conducted in the past on multi-layer and multi-track deposition using FES to study and predict the distribution of temperature, the effect of deposition process parameters on them, and the mechanism of modification of the properties of the deposition material. The following paragraphs briefly review the past work done on thermal analysis in different AM processes covering dimensionality of heat conduction, geometry of heat source, heat losses, temperature-dependent properties of deposition and substrate materials, measurement of temperature, and simulation of the deposition process.

Dai and Shaw [32] used a 3D-FE model to study the temperature field in multi-material deposition by a laser-based process and to study the effects of deposition direction and scanning rate on temperature distribution. They concluded that the temperature distribution and thermal distortion in multi-material deposition are dependent on the laser-based process and material properties. Gan et al. [33] performed FES of a plasma thermal spraying process and thermo-mechanical analysis to predict residual stresses generated in the plasma-sprayed coatings. A good agreement was observed between the simulated and experimental results. Costa et al. [34] did FES of a laser-based powder deposition process to study the effects of thermal cycles on various metallurgical phenomena involved. They also studied the effects of the substrate dimensions and idle time between the deposition of two successive layers on the microstructure and hardness of the deposition. Their result showed that the thermal properties, mechanical properties, and microstructure of the additive-manufactured component are significantly affected by the process parameters. Alimardani et al. [35] developed a transient numerical model to study the temperature distribution and thermal stresses generated during solid freeform fabrication by the laser-based process. They investigated the effects of pre-heating and clamping of the substrate to the positioning table on the temperature distribution and thermal stresses and found that pre-heating of the substrate material reduces thermal stresses due to reduced heating and cooling rates. Neela and De [36] analyzed 3D heat transfer in the laser-engineered net shaping (LENS) process to study the influence of process parameters on temperature distribution, thermal cycles, and melt pool size. Their study revealed that the selection of appropriate values of deposition process parameters results in steady dimensions of the melt pool with good inter-layer bonding. Zhao et al. [37] carried out a 3D transient heat transfer simulation of multi-layer deposition by GMA-based rapid prototyping. They investigated the effects of deposition direction on temperature distribution, thermal gradient, and thermal cycles, and also optimized deposition parameters and deposition direction through the simulation.

4.3.1 Dimensionality of Conductive Heat Transfer

Heat is transferred within the substrate material by conduction due to its interaction with the heat source during any deposition process, as shown in Figure 4.4.

Some researchers have proposed various models considering heat transfer phenomena either as one-dimensional or two-dimensional transfer within substrate material. It implies the making of the following assumptions, which lead to significant error in the prediction of temperature distribution: (i) the heat source is assumed a point source for 1D heat transfer and a line source for 2D heat transfer, (ii) the workpiece is assumed to be semi-infinite, and (iii) thermal properties of the substrate and deposition materials are assumed to be temperature-independent. To improve the accuracy of the prediction of the temperature distribution, Samarskii and Vabishchevich [39] used the following equation for 3D heat conduction during deposition.

$$\rho C_p \frac{\partial T}{\partial t} = \nabla \cdot (K \nabla T) \tag{4.1}$$

where ρ is the density (Kg/m^3), C_p is the specific heat capacity (J/Kg K), K is the thermal conductivity (W/m K), t is the time for which the heat source strikes the substrate material (sec), and T is the instantaneous temperature of the melt pool (K).

Most of the thermal processes can be numerically simulated and analyzed using heat conduction described by equation 4.1. Any additional phenomenon specific to a deposition process can be included in the boundary conditions. Chande et al. [40] introduced one of the initial numerical models for the laser welding process for conductive heat transfer, considering the temperature-dependent properties of the base material

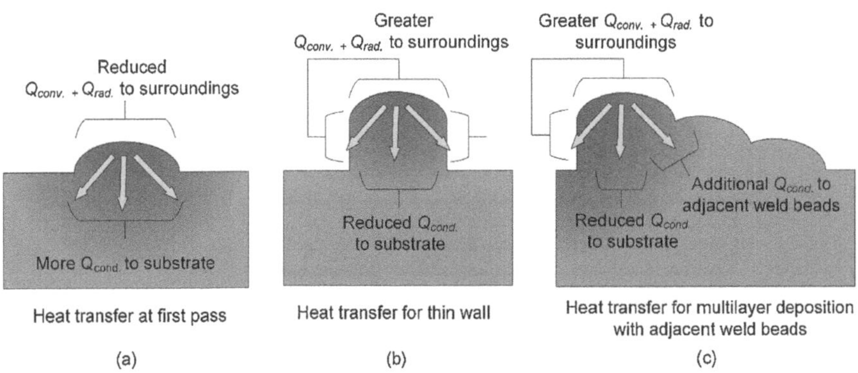

FIGURE 4.4 Schematic representation of conductive heat transfer during an AM process: (a) for single-track deposition, (b) for thin-walled structures, and (c) for overlapping depositions [38]. AM, Additive manufacturing.

and its latent heat of fusion. Kim and Basu [41] used a 3D heat conduction model to predict deposition geometry characteristics and temperature distribution during the GMA-based deposition process. The same model was further used by Kim and Na [42] to study fluid flow in the weld pool formed by a pulsed gas tungsten arc (GTA)-based deposition process. Fan et al. [43] numerically analyzed the conductive heat transfer and fluid flow considering electromagnetic force and Marangoni flow on a stationary melt pool formed in the substrate material by a GTA-based deposition process.

4.3.2 Geometry of Heat Source

The geometry of the heat source used in the thermal modeling and FES of an AM process should be as close as possible to the actual heat source and should give value of heat flow rate per unit volume approximately equal to its actual value. Rosenthal [44,45] developed the following equation for a moving line heat source to predict temperature distribution in the arc welding process:

$$T - T_i = \frac{\rho C V_e Q}{8\pi r K^2}(r+\xi) \tag{4.2}$$

where T_i is the ambient temperature (K); Q is the heat input (W); r is the distance from the center of the heat source (m); K, ρ, and C are the thermal conductivity (W/m K), density (Kg/m^3), and specific heat capacity (J/Kg K) of the base material; V_e is the velocity of the heat source (m/sec); and ξ is the distance traveled by the heat source (m). Pavelic et al. [46] proposed a model for a disc-shaped heat source as shown in Figure 4.5. This heat source tries to overcome some limitations of the point and line heat source model of Rosenthal.

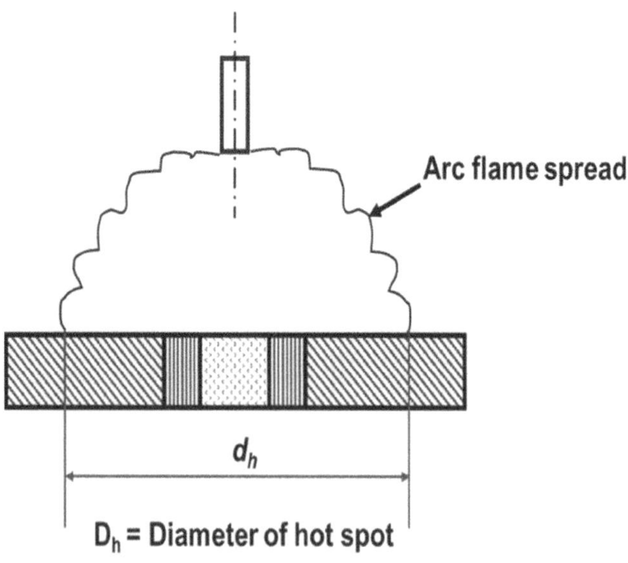

FIGURE 4.5 Schematic of a disc-shaped heat source.

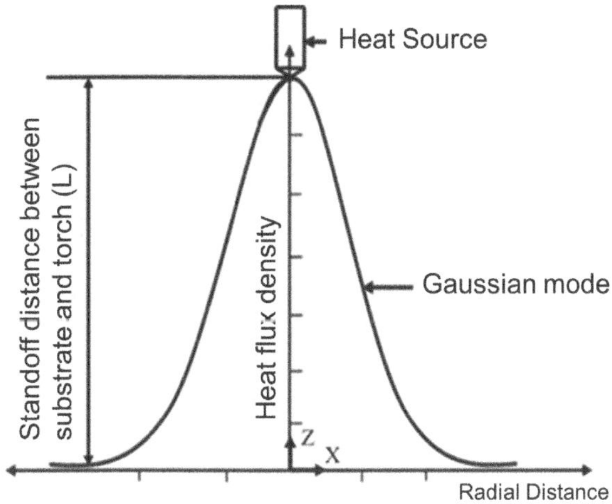

FIGURE 4.6 Distribution of heat flux density for a Gaussian heat source.

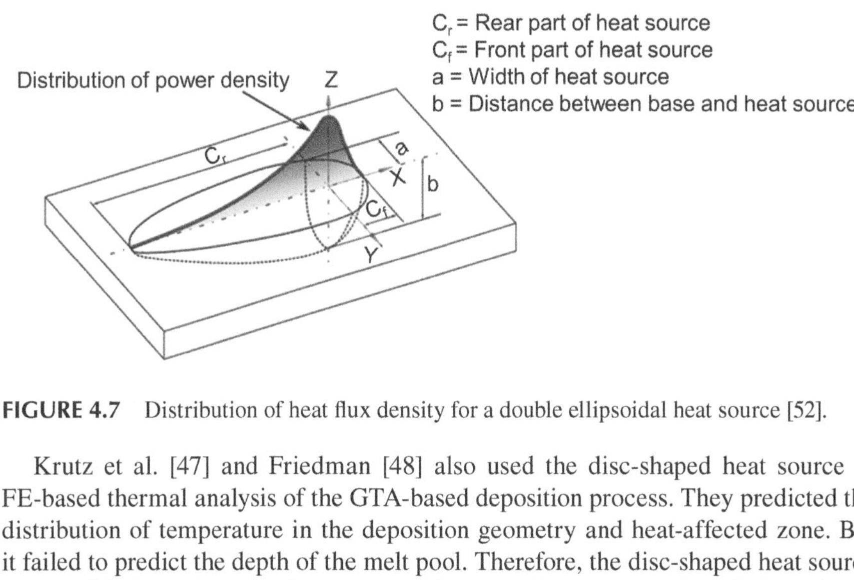

FIGURE 4.7 Distribution of heat flux density for a double ellipsoidal heat source [52].

Krutz et al. [47] and Friedman [48] also used the disc-shaped heat source in FE-based thermal analysis of the GTA-based deposition process. They predicted the distribution of temperature in the deposition geometry and heat-affected zone. But it failed to predict the depth of the melt pool. Therefore, the disc-shaped heat source was modified as a Gaussian heat source to improve the accuracy of prediction of temperature distribution and thermal cycles. Jeong and Cho [49] carried out a transient FES considering a Gaussian heat source to calculate temperature distribution during deposition of Ti-6Al-4V on a substrate of the same material by the GTA-based deposition process. Figure 4.6 shows the distribution of heat flux density for a Gaussian heat source.

Goldak et al. [50] proposed a double ellipsoidal heat source by modifying the Gaussian heat source to consider an unsymmetrical distribution of heat within the substrate material. Figure 4.7 shows the distribution of heat flux density for the double ellipsoidal heat source. Nguyen et al. [51] used this heat source for a semi-infinite body to predict bead geometry parameters, temperature distribution, thermal cycles,

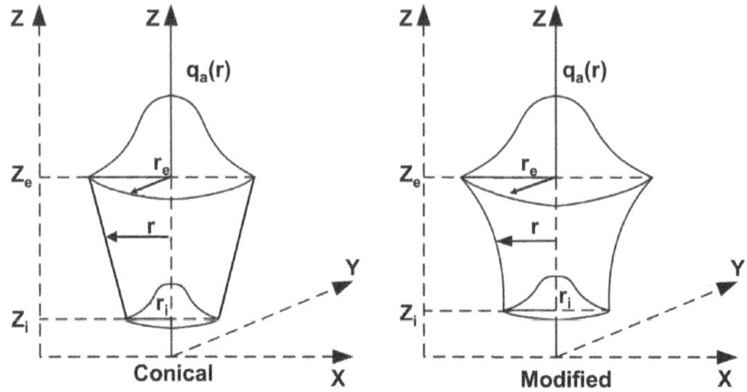

FIGURE 4.8 Schematic of a conical-shaped heat source; and a modified conical-shaped heat source [54].

and residual stresses generated during the submerged gas metal arc-based deposition process.

Researchers have developed hybrid heat sources by combining a conical heat source (Figure 4.8(a)) with a Gaussian heat source and referred to it as the modified conical heat source model (Figure 4.8(b)). Luo et al. [53] did the thermal analysis using a modified conical heat source for deposition of magnesium alloy on a substrate of similar material by the electron beam melting process.

4.3.3 Heat Losses

Convective and radiative heat losses from the substrate and deposition materials to the surroundings take place during any AM process. Therefore, they have to be incorporated in the modeling, simulation, prediction, and optimization to increase their accuracy. Convective heat loss per unit area 'q_c' (W/m^2) and radiative heat loss per unit area 'q_r' (W/m^2) from the substrate (or deposition) material are expressed by equations 4.3 and 4.4:

$$q_c = h_{\text{conv}} \left(T_m - T_i\right) \tag{4.3}$$

$$q_r = \varepsilon \sigma_{\text{sbc}} \left(T_m^4 - T_i^4\right) \tag{4.4}$$

where h_{conv} is the convective heat transfer coefficient of the substrate (or deposition) material (W/m^2 K), σ_{sbc} is the Stefan-Boltzmann constant (W/m^2 K^4), ε is the emissivity of the substrate (or deposition) material, T_m is the melting temperature of the substrate (or deposition) material (K), and T_i is the ambient temperature (K). Abid and Siddique [55] combined the heat losses due to convection and radiation in the following single relation and referred to the combined heat loss as 'q_{cr}' (W/m^2 K) for the GMA welding process:

$$q_{\text{cr}} = \frac{\varepsilon \sigma_{\text{sbc}} \left(T_{\text{ms}}^4 - T_i^4\right)}{\left(T_{\text{ms}} - T_i\right)} + h_{\text{conv}} \tag{4.5}$$

4.3.4 Temperature-Dependent Properties of the Materials

Some of the developed analytical models have neglected the temperature dependence of the properties of the deposition and substrate materials. However, it adversely affects the accuracy of the modeling and 3D-FES of melt pool formation, temperature distribution, thermal cycles, residual stresses, and distortion during the deposition process. Zacharia et al. [56] simulated the influence of temperature-dependent properties of the substrate and deposition materials on the surface tension of the melt pool in laser and GTA-based deposition processes. Simulated results were found to be in good agreement with the experimental results. Mundra et al. [57] studied the influence of the thermal and physical properties (namely viscosity, thermal diffusivity, surface tension, and coefficient of energy absorption) of the substrate and deposition materials on the diameter of the melt pool, and the penetration depth in an arc-based deposition process.

4.3.5 Simulation of Deposition Process

The most challenging task in an AM process is to simulate the deposition of the material with time. Matsumoto et al. [58] carried out 2D-FES for the single-layer deposition of nickel-based powder, simulating the actual flow of the powder with carrier gas and analyzed its effect on the deposition geometry. Shiomi et al. [59] used the element birth and death technique to simulate deposition of chrome-molybdenum-steel powder by the selective laser sintering (SLS) process. Mercelis and Kruth [60] used FES to predict the distribution of the residual stresses after deposition of each layer of a metallic powder on a steel substrate by the SLS process. They did this up to fifty layers of deposition.

4.3.6 Measurement of Temperature

Validation of the predicted results by FES requires their comparison with the experimentally measured thermal cycles. Instruments such as thermocouples, high-speed camera, thermal imaging camera, and optical pyrometer have been used for the measurement of temperature and thermal cycles during AM processes. Hofmeister et al. [61] used a charge-coupled device (CCD) camera to monitor the temperature distribution in the melt pool formed during the LENS process. On the contrary, Zeng et al. [62] monitored the temperature distribution in the melt pool using a pyrometer in the LENS process. Schwerdtfeger et al. [63] used an infrared (IR) camera to measure temperature and capture the flaws, porosities, and imperfections in the layers deposited by the electron beam melting process. Heigel et al. [64] carried out in-situ measurement of temperature using K-type thermocouples fixed at several locations on the substrate to validate the results of FES in the laser-based AM process.

4.4 SIMULATION OF RESIDUAL STRESSES IN THIN-WALLED STRUCTURES

FES is essential for predicting temperature distribution, thermal cycles, thermal distortion, and residual stress generated during AM deposition, thereby reducing the need for expensive, time- and resource-intensive experimentation.

4.4.1 Simulation of Residual Stresses

Chew et al. [65] used FES to simulate the residual stresses generated in the deposition of AISI 4340 steel powder on a substrate of similar material by the laser cladding process. The simulated values were validated by comparing them with the residual stresses measured by X-ray diffraction. Foroozmehr et al. [66] used 3D-FES to predict thermal cycles and temperature distribution in the deposition of 316L stainless steel by the selective laser melting process using the concept of element birth and death. Denlinger et al. [67] performed FE-based modeling to predict the residual stresses and distortion in the deposition of Ti-6Al-4V alloy by the electron beam direct manufacturing process. The simulated results were validated by comparing them with the experimental values measured by the strain gauges. Hemmesi et al. [68] employed a finite element approach to calculate the residual stresses formed during GTA welding of S355J2H steel. X-ray diffraction and neutron diffraction techniques were used to determine experimental values of the residual stresses, and the calculated and experimental values were found to be in good agreement. Ding et al. [69] developed a transient thermo-elastic-plastic model to study the temperature, residual stresses, and deformation generated in the substrate and multi-layer metallic deposition by wire- and arc-based AM. The simulated residual stresses were compared with their experimental values obtained by neutron diffraction technique, and a quite good agreement was found between them.

4.4.2 Measurement of Residual Stresses

Measurement of residual stresses induced during any AM process is done either during or after completion of the deposition process using either destructive or non-destructive methods. Olabi and Hashmi [70] used the hole-drilling method to measure residual stress distribution generated during the plasma transferred arc (PTA)-based deposition process. This is a destructive method where the rosette strain gauge, as shown in Figure 4.9, was fixed in a small hole drilled at the location where the residual stresses are to be measured. Zhu and Chao [71] used a dial gauge to measure the distortion caused in the substrate material during continuous movement of GMA. Zheng et al. [72] used the XRD method to determine the residual stresses in thin films prepared by a pulsed-laser deposition process. An et al. [73] used the neutron diffraction technique to measure the residual stresses formed in a curved thin-walled metallic structure manufactured by the SLS process.

4.5 SUMMARY

The research work presented in this chapter comprehensively explores the various facets of AM, particularly focusing on the modeling, simulation, and thermal analysis of deposition processes, with an emphasis on metallic materials and complex structures. Additive manufacturing, also known as AM, layered manufacturing, or solid freeform fabrication, is a bottom-up approach that allows for the creation of complex geometries, customization, repair, and refurbishment of components with minimal material waste. Unlike conventional subtractive or deformative manufacturing, AM

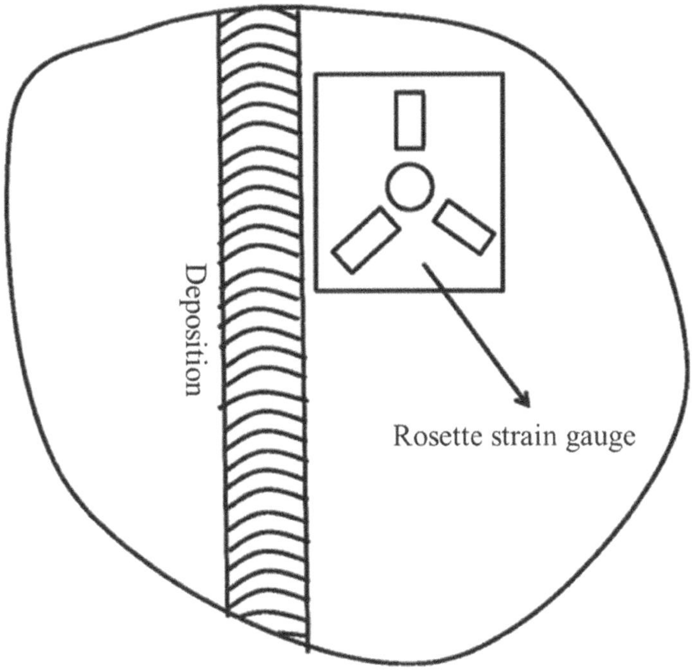

FIGURE 4.9 Schematic of the hole-drilling method for measurement of residual stresses [70].

offers significant advantages in material efficiency, energy consumption, and environmental impact.

The chapter begins by categorizing the deposition patterns used in AM processes, namely single-layer single-track, single-layer multi-track, multi-layer single-track, and multi-layer multi-track. These patterns are foundational to various applications, such as coatings, cladding, remanufacturing, and fabrication of thin-walled structures. However, applying AM to high-melting-point metals, ceramics, and biocompatible materials remains a challenge due to limitations in existing processes. To overcome this, research has focused on developing new AM methods, enhancing existing processes, and using modeling and optimization to improve outcomes.

A central theme in this work is the importance of deposition geometry characteristics. The accuracy and quality of AM components heavily depend on the precise geometry of single-layer single-track depositions, including parameters such as height, width, aspect ratio, dilution, and root angle. These parameters influence bonding strength between layers, dimensional accuracy, and overall structural integrity. The melt pool dynamics comprising zones such as the melt pool, solid–liquid interface, and solid substrate are also critical. Accurate modeling of these zones is essential to predict and control the deposition geometry.

The research delves into three primary approaches for modeling deposition geometry: thermal modeling, empirical modeling, and FES. Thermal models use mathematical equations to predict melt pool dimensions based on input parameters like laser power, travel speed, and powder feed rate. These models range from simple elliptical or hemispherical approximations to more complex formulations that consider surface tension and powder flow. Empirical modeling uses statistical techniques, such as regression analysis and ANNs, to establish relationships between process parameters and deposition outcomes. While empirical models are useful, their accuracy is limited by the quantity and quality of experimental data. FES offers a more comprehensive and dynamic approach by simulating the physical phenomena involved in AM, including heat transfer, fluid flow, phase transformations, and residual stress formation.

The chapter provides an in-depth review of prior work in thermal modeling and FES. Researchers have developed both 2D and 3D thermal models to study heat flow and melt pool formation. Factors such as the geometry of the heat source, thermal conductivity, and Marangoni flow significantly affect simulation outcomes. Different heat source geometries, including point, line, disc-shaped, Gaussian, double ellipsoidal, and modified conical shapes, have been used to better represent the actual heat distribution during AM. Modeling must also account for heat losses due to convection and radiation, which influence melt pool temperature and cooling rates. Some models have incorporated temperature-dependent material properties to improve prediction accuracy.

The research further investigates thermal analysis of thin-walled structures. It underscores the necessity of understanding temperature distribution, gradients, and thermal cycles during AM to minimize distortion and residual stresses. FES helps identify optimal heat input and process parameters, thereby reducing experimental workload and cost. Studies on multi-layer, multi-track deposition processes have shown how simulation can predict temperature fields, optimize scanning strategies, and improve mechanical properties of the final component.

Dimensionality of heat conduction, whether modeled in 1D, 2D, or 3D, plays a significant role in simulation accuracy. While simpler models assume idealized conditions, 3D models incorporating temperature-dependent properties provide better predictions. The heat source's geometry is vital to modeling accuracy; earlier models by Rosenthal, Pavelic, and Goldak laid the groundwork for modern simulations by introducing moving point sources and Gaussian or double ellipsoidal distributions. These have been refined to simulate real-time deposition processes more closely.

To validate these models, accurate temperature measurements during AM are essential. Techniques such as thermocouples, high-speed cameras, IR imaging, and optical pyrometers are commonly used. These instruments capture temperature profiles, thermal cycles, and help detect imperfections like porosity or cracks. For example, CCD cameras and pyrometers have been used in LENS processes, while IR cameras are employed in electron beam melting.

The final section of the chapter addresses simulation and measurement of residual stresses in thin-walled structures. Residual stress is a major concern in AM due to repeated heating and cooling cycles. Simulations are used to predict these stresses and mitigate their impact by optimizing process parameters. Methods such as transient thermo-elastic-plastic modeling and element birth-death techniques simulate material addition and heat-induced stress formation over time. The simulations are validated through experimental techniques like X-ray diffraction, neutron diffraction, and hole-drilling methods. These techniques provide reliable data on stress distribution and are essential for comparing and validating simulation results.

In conclusion, the chapter provides a holistic view of the critical role modeling and simulation play in the field of AM, especially for metallic materials and complex geometries. It highlights the evolution from basic thermal models to advanced FES techniques that consider temperature-dependent properties, heat loss mechanisms, and actual heat source geometries. Empirical models and simulation tools complement each other in improving process predictability and reliability. Thermal analysis and residual stress simulation are crucial for designing robust AM components with minimal defects. Measurement techniques for validating these simulations are equally important for bridging the gap between theoretical predictions and real-world outcomes. This body of work not only consolidates existing research but also serves as a guide for future innovations aimed at expanding AM's applicability to more challenging materials and critical industrial sectors.

REFERENCES

1. Alcisto, J., Enriquez, A., Garcia, H., Hinkson, S., Steelman, T., Silverman, E., Valdovino, P., Gigerenzer, H., Foyos, J., Ogren, J., Dorey, J., Karg, K., McDonald, T., Es-Said, O. S., 2011. Tensile properties and microstructures of laser-formed Ti-6Al-4V. *Journal of Materials Engineering and Performance* 20(2), 203–212.
2. Liu, W., DuPont, J. N., 2003. Fabrication of functionally graded TiC/Ti composites by laser engineered net shaping. *Scripta Materialia* 48(9), 1337–1342. https://doi.org/10.1016/S1359-6462(03)00020-4
3. Ahsan, M. N., Paul, C. P., Kukreja, L. M., Pinkerton, A. J., 2011. Porous structures fabrication by continuous and pulsed laser metal deposition for biomedical applications: Modelling and experimental investigation. *Journal of Materials Processing Technology* 211(4), 602–609. https://doi.org/10.1016/j.jmatprotec.2010.11.014
4. Kumar, P., Jain, N. K., Sawant, M. S., 2022. Development of theoretical models for dimensions of single-layer multi-track and multi-layer multi-track depositions by µ-PTA additive manufacturing process. *Journal of Materials Research and Technology* 17, 95–110. https://doi.org/10.1016/j.jmrt.2021.12.124
5. Wohler, T., 2009. State of the Industry - Annual Worldwide Progress Report. *Vasa*, 1–8.
6. Pinkerton, A. J., Li, L., 2004. An analytical model of energy distribution in laser direct metal deposition. *Proceedings of the Institution of Mechanical Engineers Part B: Journal of Engineering Manufacture* 218, 363–374.
7. Lakshminarayanan, A. K., Balasubramanian, V., Varahamoorthy, R., Babu, S., 2008. Predicting the dilution of plasma transferred arc hardfacing of stellite on carbon steel using response surface methodology. *Metals and Materials International* 14(6), 779–789. https://doi.org/10.1007/BF03027996
8. Nikam, S. H., Jain, N. K., Jhavar, S., 2016. Thermal modeling of geometry of single-track deposition in micro-plasma transferred arc deposition process. *Journal of Materials Processing Technology* 230, 121–130. https://doi.org/10.1016/j.jmatprotec.2015.11.022
9. Bamberger, M., Kaplan, W. D., Medres, B., Shepeleva, L., 1998. Calculation of process parameters for laser alloying and cladding. *Journal of Laser Applications* 10(1), 29–33. https://doi.org/10.2351/1.521829
10. Picasso, M., Hoadley, A., 1994. Finite element simulation of laser surface treatments including convection in the melt pool. *International Journal of Numerical Methods for Heat and Fluid Flow* 4, 61–83.

11. Dykhuizen, R. C., Dobranich, D., 1998. *Scaling analysis for the LENS process*. Sandia National Laboratories Internal Report.
12. Liu, J., Li, J., 2006. Direct fabrication of thin-wall parts by laser cladding. *Journal of Manufacturing Processes* 8(1), 1–7. https://doi.org/10.1016/S1526-6125(06)70096-2
13. Lalas, C., Tsirbas, K., Salonitis, K., Chryssolouris, G., 2007. An analytical model of the laser clad geometry. *International Journal of Advanced Manufacturing Technology* 32(1–2), 34–41. https://doi.org/10.1007/s00170-005-0318-0
14. Hua, T., Jing, C., Xin, L., Fengying, Z., Weidong, H., 2008. Research on molten pool temperature in the process of laser rapid forming. *Journal of Materials Processing Technology* 198(1–3), 454–462. https://doi.org/10.1016/j.jmatprotec.2007.06.090
15. Cheikh, H. El., Courant, B., Hascoët, J.-Y., Guillén, R., 2012. Prediction and analytical description of the single laser track geometry in direct laser fabrication from process parameters and energy balance reasoning. *Journal of Materials Processing Technology* 212(9), 1832–1839. https://doi.org/10.1016/j.jmatprotec.2012.03.016
16. Ya, W., Pathiraj, B., Liu, S., 2016. 2D modelling of clad geometry and resulting thermal cycles during laser cladding. *Journal of Materials Processing Technology* 230, 217–232. https://doi.org/10.1016/j.jmatprotec.2015.11.012
17. Nagesh, D. S., Datta, G. L., 2002. Prediction of weld bead geometry and penetration in shielded metal-arc welding using artificial neural networks. *Journal of Materials Processing Technology* 123(2), 303–312.https://doi.org/10.1016/S0924-0136(02)00101-2
18. Palani, P. K., Murugan, N., 2006. Development of mathematical models for prediction of weld bead geometry in cladding by flux cored arc welding. *International Journal of Advanced Manufacturing Technology* 30(7–8), 669–676. https://doi.org/10.1007/s00170-005-0101-2
19. Onwubolu, G. C., Davim, J. P., Oliveira, C., Cardoso, A., 2007. Prediction of clad angle in laser cladding by powder using response surface methodology and scatter search. *Optics and Laser Technology* 39(6), 1130–1134. https://doi.org/10.1016/j.optlastec.2006.09.008
20. Paulo Davim, J., Oliveira, C., Cardoso, A., 2008. Predicting the geometric form of clad in laser cladding by powder using multiple regression analysis (MRA). *Materials and Design* 29(2), 554–557. https://doi.org/10.1016/j.matdes.2007.01.023
21. Sun, Y., Hao, M., 2012. Statistical analysis and optimization of process parameters in Ti6Al4V laser cladding using Nd:YAG laser. *Optics and Lasers in Engineering* 50(7), 985–995. https://doi.org/10.1016/j.optlaseng.2012.01.018
22. Nenadl, O., Ocelík, V., Palavra, A., De Hosson, J. T. M., 2014. The prediction of coating geometry from main processing parameters in laser cladding. *Physics Procedia* 56(C), 220–227. https://doi.org/10.1016/j.phpro.2014.08.166
23. Xiong, J., Zhang, G., Hu, J., Wu, L., 2014. Bead geometry prediction for robotic GMAW-based rapid manufacturing through a neural network and a second-order regression analysis. *Journal of Intelligent Manufacturing* 25(1), 157–163. https://doi.org/10.1007/s10845-012-0682-1
24. Saqiba, S., Urbanica, R. J., Aggarwala K., 2014. Analysis of laser cladding bead morphology for developing additive manufacturing travel paths. *Procedia CIRP* 17, 824–829.
25. Lee, E. M., Shin, G. Y., Yoon, H. S., Shim, D. S., 2017. Study of the effects of process parameters on deposited single track of M4 powder based direct energy deposition. *Journal of Mechanical Science and Technology* 31(7), 3411–3418. https://doi.org/10.1007/s12206-017-0239-5
26. Nabhani, M., Razavi, R. S., Barekat, M., 2018. An empirical-statistical model for laser cladding of Ti-6Al-4V powder on Ti-6Al-4V substrate. *Optics and Laser Technology* 100, 265–271.

27. Hoadley, A. F. A., Rappaz, M., Zimmermann, M., 1991. Heat-flow simulation of laser remelting with experimenting validation. *Metallurgical Transactions B* 22(1), 101–109. https://doi.org/10.1007/BF02672531
28. Kim, J. D., Peng, Y., 2000. Melt pool shape and dilution of laser cladding with wire feeding. *Journal of Materials Processing Technology* 104(3), 284–293. https://doi.org/10.1016/S0924-0136(00)00528-8
29. Lei, Y. P., Murakawa, H., Shi, Y. W., Li, X. Y., 2001. Numerical analysis of the competitive influence of Marangoni flow and evaporation on heat surface temperature and molten pool shape in laser surface remelting. *Computational Materials Science* 21(3), 276–290. https://doi.org/10.1016/S0927-0256(01)00143
30. Toyserkani, E., Khajepour, A., Corbin, S., 2003. Three-dimensional finite element modeling of laser cladding by powder injection: Effects of powder feed rate and travel speed on the process. *Journal of Laser Applications* 15(3), 153–160. https://doi.org/10.2351/1.1585087
31. Han, L., Phatak, K. M., Liou, F. W., 2004. Modeling of laser cladding with powder injection. *Metallurgical and Materials Transactions B* 35(6), 1139–1150. https://doi.org/10.1007/s11663-004-0070-0
32. Dai, K., Shaw, L., 2001. Thermal and stress modeling of multi-material laser processing. *Acta Materialia* 49(20), 4171–4181. https://doi.org/10.1016/S1359-6454(01)00312-3
33. Gan, Z., Ng, H. W., Devasenapathi, A., 2004. Deposition-induced residual stresses in plasma-sprayed coatings. *Surface and Coatings Technology* 187(2–3), 307–319. https://doi.org/10.1016/j.surfcoat.2004.02.010
34. Costa, L., Vilar, R., Reti, T., Deus, A. M., 2005. Rapid tooling by laser powder deposition: Process simulation using finite element analysis. *Acta Materialia* 53(14), 3987–3999. https://doi.org/10.1016/j.actamat.2005.05.003
35. Alimardani, M., Toyserkani, E., Huissoon, J. P., 2007. A 3D dynamic numerical approach for temperature and thermal stress distributions in multilayer laser solid freeform fabrication process. *Optics and Lasers in Engineering* 45(12), 1115–1130. https://doi.org/10.1016/j.optlaseng.2007.06.010
36. Neela, V., De, A., 2009. Three-dimensional heat transfer analysis of LENSTM process using finite element method. *International Journal of Advanced Manufacturing Technology* 45(9–10), 935–943. https://doi.org/10.1007/s00170-009-2024-9
37. Zhao, H., Zhang, G., Yin, Z., Wu, L., 2011. A 3D dynamic analysis of thermal behavior during single-pass multi-layer weld-based rapid prototyping. *Journal of Materials Processing Technology* 211(3), 488–495. https://doi.org/10.1016/j.jmatprotec.2010.11.002
38. Cunningham, C. R., Flynn, J. M., Shokrani, A., Dhokia, V., Newman, S. T., 2018. Strategies and processes for high quality wire arc additive manufacturing. *Additive Manufacturing* 22, 672–686. https://doi.org/10.1016/j.addma.2018.06.020
39. Samarskii, A. A., Vabishchevich, P. N., 1995. *Computational heat transfer, Mathematical modelling.* Chichester John Wiley & Sons, Hoboken, New Jersey.
40. Chande, T., Mazumder, J., 1984. Estimating effects of processing conditions and variable properties upon pool shape, cooling rates, and absorption-coefficient in laser-welding. *Journal of Applied Physics* 56(7), 1981–1986.
41. Kim, I., Basu, A., 1998. A mathematical model of heat transfer and fluid flow in the gas metal arc welding process. *Journal of Materials Processing Technology* 77, 17–24. https://doi.org/10.1016/S0924-0136(97)00383-X
42. Kim, W. H., Na, S. J., 1998. Heat and fluid flow in pulsed current GTA weld pool. *International Journal of Heat and Mass Transfer* 41(21), 3213–3227. https://doi.org/10.1016/S0017-9310(98)00052-0
43. Fan, H. G., Tsai, H. L., Na, S. J., 2000. Heat transfer and fluid flow in a partially or fully penetrated weld pool in gas tungsten arc welding. *International Journal of Heat and Mass Transfer* 44, 417–428. https://doi.org/10.1016/S0017-9310(00)00094-6

44. Rosenthal, D., 1941. Mathematical theory of heat distribution during welding and cutting. *Welding Journal*, 20(5), 220–234.
45. Rosenthal, D., 1946. The theory of moving sources of heat and its application to metal treatments. *Transactions of ASME* 43(11), 849–865.
46. Pavelic, V., Tanbakuchi, O., Uyehara, O. A., Myers, P. S., 1969. Experimental and computed temperature histories in gas tungsten-arc welding of thin plates. *Welding Journal* 48(7), 295–305.
47. Krutz, G. W., Segerlind, L. J., 1978. Finite element analysis of welded structures. *Welding Journal Research* 57(Supplement), 211–216.
48. Friedman, E., 1978. Analysis of Weld Puddle Distortion and Its Effect on Penetration. *Welding Journal* 56(6), 161–166.
49. Jeong, S. K., Cho, H. S., 1997. An analytical solution to predict the transient temperature distribution in fillet arc welds. *Welding Journal-Including Welding Research Supplement* 76(6), 223.
50. Goldak, J., Chakravarti, A., Bibby, M., 1984. A new finite element model for welding heat sources. *Metallurgical Transactions B* 15(2), 299–305. https://doi.org/10.1007/BF02667333
51. Nguyen, N. T., Ohta, A., Matsuoka, K., Suzuki, N., Maeda, Y., 1999. Analytical solutions for transient temperature of semi-infinite body subjected to 3-D moving heat sources. *Welding Research Supplement* 1, 265–274.
52. Fritschle, T., Kaess, M., Weihe, S., Werz, M. 2025. Investigation of the thermo-mechanical modeling of the manufacturing of large-scale wire arc additive manufacturing components with an outlook towards industrial applications. *Journal of Manufacturing and Materials Processing* 9(5), 166. https://doi.org/10.3390/jmmp9050166
53. Luo, Y., You, G., Ye, H., Liu, J., 2010. Simulation on welding thermal effect of AZ61 magnesium alloy based on three-dimensional modeling of vacuum electron beam welding heat source. *Vacuum* 84(7), 890–895. https://doi.org/10.1016/j.vacuum.2009.12.005
54. Wang, H. X., Wei, Y. H., Yang, C. L., 2007. Numerical simulation of variable polarity vertical-up plasma arc welding process. *Computational Materials Science* 38(4), 571–587. https://doi.org/10.1016/j.commatsci.2006.03.017
55. Abid, M., Siddique, M., 2005. Numerical simulation to study the effect of tack welds and root gap on welding deformations and residual stresses of a pipe-flange joint. *International Journal of Pressure Vessels and Piping* 82(11), 860–871. https://doi.org/10.1016/j.ijpvp.2005.06.008
56. Zacharia, T., David, S. A., Vitek, J. M., Debroy, T., 1989. Heat transfer during Nd: Yag pulsed laser welding and its effect on solidification structure of austenitic stainless steels. *Metallurgical Transactions A* 20(5), 957–967. https://doi.org/10.1007/BF02651661
57. Mundra, K., Debroy, T., Zacharia, T., David, S. A., 1992. Role of thermophysical properties in weld pool modeling. *Welding Journal* 71(9), S313–S320.
58. Matsumoto, M., Shiomi, M., Osakada, K., Abe, F., 2002. Finite element analysis of single layer forming on metallic powder bed in rapid prototyping by selective laser processing. *International Journal of Machine Tools and Manufacture* 42(1), 61–67. https://doi.org/10.1016/S0890-6955(01)00093-1
59. Shiomi, M., Osakada, K., Nakamura, K., Yamashita, T., Abe, F., 2004. Residual stress within metallic model made by selective laser melting process. *CIRP Annals - Manufacturing Technology* 53(1), 195–198. https://doi.org/10.1016/S0007-8506(07)60677-5
60. Mercelis, P., Kruth, J., 2006. Residual stresses in selective laser sintering and selective laser melting. *Rapid Prototyping Journal* 12(5), 254–265. https://doi.org/10.1108/13552540610707013

61. Hofmeister, W., Knorovsky, G. A., Maccallum, D. O., 1999. Video monitoring and control of the LENS process. *9th International Conference of Computer Technology in Welding by American Welding Society.*
62. Zeng, K., Pal, D., Stucker, B. E., 2012. A review of thermal analysis methods in laser sintering and selective laser melting. *Proceedings of the Solid Freeform Fabrication Symposium*, 796–814.
63. Schwerdtfeger, J., Singer, R. F., Körner, C., 2012. In situ flaw detection by IR-imaging during electron beam melting. *Rapid Prototyping Journal* 18(4), 259–263. https://doi.org/10.1108/13552541211231572
64. Heigel, J. C. C., Michaleris, P., Reutzel, E. W. W., 2015. Thermo-mechanical model development and validation of directed energy deposition additive manufacturing of Ti-6Al-4V. *Additive Manufacturing* 5, 9–19. https://doi.org/10.1016/j.addma.2014.10.003
65. Chew, Y., Pang, J. H. L., Bi, G., Song, B., 2015. Thermo-mechanical model for simulating laser cladding induced residual stresses with single and multiple clad beads. *Journal of Materials Processing Technology* 224, 89–101. https://doi.org/10.1016/j.jmatprotec.2015.04.031
66. Foroozmehr, A., Badrossamay, M., Foroozmehr, E., Golabi, S., 2016. Finite element simulation of selective laser melting process considering optical Penetration Depth of laser in powder bed. *Materials and Design* 89, 255–263. https://doi.org/10.1016/j.matdes.2015.10.002
67. Denlinger, E. R., Heigel, J. C., Michaleris, P., 2015. Residual stress and distortion modeling of electron beam direct manufacturing Ti-6Al-4V. *Proceedings of the Institution of Mechanical Engineers, Part B: Journal of Engineering Manufacture* 229(10), 1803–1813. https://doi.org/10.1177/0954405414539494
68. Hemmesi, K., Farajian, M., Boin, M., 2017. Numerical studies of welding residual stresses in tubular joints and experimental validations by means of x-ray and neutron diffraction analysis. *Materials and Design* 126(4), 339–350. https://doi.org/10.1016/j.matdes.2017.03.088
69. Ding, J., Colegrove, P., Mehnen, J., Ganguly, S., Almeida, P. M. S., Wang, F., Williams, S. 2011. Thermo-mechanical analysis of wire and arc additive layer manufacturing process on large multi-layer parts. *Computational Materials Science* 50(12), 3315–3322. https://doi.org/10.1016/j.commatsci.2011.06.023
70. Rossini, N. S., Dassisti, M., Benyounis, K. Y., Olabi, A. G. 2012. Methods of measuring residual stresses in components. *Materials & Design* 35, 572–588. https://doi.org/10.1016/j.matdes.2011.08.022
71. Zhu, X. K., Chao, Y. J., 2002. Effects of temperature-dependent material properties on welding simulation. *Computers and Structures* 80(11), 967–976.
72. Zheng, X., Li, J., Zhou, Y., 2004. X-ray diffraction measurement of residual stress in PZT thin films prepared by pulsed laser deposition. *Acta Materialia* 52(11), 3313–3322. https://doi.org/10.1016/j.actamat.2004.02.047
73. An, K., Yuan, L., Dial, L., Spinelli, I., Stoica, A. D., Gao, Y., 2017. Neutron residual stress measurement and numerical modeling in a curved thin-walled structure by laser powder bed fusion additive manufacturing. *Materials and Design* 135, 122–132. https://doi.org/10.1016/j.matdes.2017.09.018

5 Data-Driven Models for Feature Measurement and Defects Monitoring in Additive Manufacturing

Deepika Nikam, Akriti Nigam, and Tejaswini Bhosale

5.1 INTRODUCTION

Additive manufacturing (AM) has revolutionized modern production systems by enabling the fabrication of complex, custom-designed parts with reduced material waste, lower production time, and minimal tooling requirements. Among the various applications of AM, the manufacturing of thin-walled structures has gained increasing importance due to their prevalence in aerospace, biomedical, defense, and automotive industries. These structures are particularly valued for their high strength-to-weight ratios and functional design flexibility. However, thin-walled components are highly sensitive to process-induced defects due to their geometric fragility, low thermal mass, and complex heat transfer behavior. Defects such as porosity, cracking, lack of fusion (LOF), delamination, and warping can significantly degrade the mechanical performance, surface integrity, and reliability of AM-produced parts [1, 2].

Conventional approaches for defect detection in AM rely on post-process inspection techniques such as optical microscopy, X-ray computed tomography (CT), ultrasonic testing, and coordinate measuring machines. While these methods can provide detailed characterization, they are inherently offline, time-consuming, and costly. Moreover, they are not scalable for high-throughput production environments or real-time monitoring needs. Manual inspection introduces additional subjectivity, human error, and inconsistency in quality assurance. As the industrial adoption of AM scales up, especially within Industry 4.0 paradigms, there is a growing demand for intelligent, automated, and adaptive monitoring systems that can detect and even predict defects during the manufacturing process itself [3, 4].

 DOI: 10.1201/9781003571209-5

This demand is being met by the increasing integration of sensors and real-time data acquisition systems within AM platforms. Thermal cameras, acoustic sensors, melt pool monitoring units, and high-resolution vision systems generate vast volumes of real-time data during the build process. These multimodal datasets offer valuable insights into the process health, enabling early detection of anomalies that may result in structural defects. The challenge, however, lies in the effective interpretation and utilization of this data. Traditional rule-based systems often fail to scale or adapt to new materials, geometries, or process parameters. As a result, the focus has shifted to data-driven models, particularly those based on machine learning (ML) and deep learning (DL) [5, 6].

ML methods have shown significant promise in defect classification and process parameter prediction in AM. Algorithms such as decision trees, support vector machines (SVM), random forests, and k-nearest neighbors are capable of learning from labeled datasets to distinguish between defective and non-defective builds. These models are used to analyze features derived from sensor data, such as temperature profiles, acoustic signatures, or visual patterns, to predict the likelihood of defect formation [7, 8]. While ML models are interpretable and relatively lightweight, their performance often depends heavily on feature engineering and expert-defined thresholds.

In contrast, DL models have the capacity to automatically extract meaningful features from raw and unstructured data such as images, time-series signals, and audio patterns. Convolutional neural networks (CNNs) have emerged as powerful tools for analyzing thermal images, optical scans, and high-speed videos of the AM process. These models can detect surface cracks, spatter, porosity, and irregularities in melt pool dynamics with high accuracy [9, 10]. Recurrent neural networks (RNNs), particularly long short-term memory (LSTM) networks, are suited for capturing temporal patterns in sensor data, making them useful for detecting time-evolving anomalies during layer-by-layer deposition [11]. Autoencoders, including their variational forms, are increasingly being used for unsupervised anomaly detection by learning compact representations of normal process behavior and flagging deviations as potential defects [12].

The introduction of transfer learning (TL) has further accelerated the deployment of DL models in AM, allowing the reuse of pre-trained models such as VGGNet, ResNet, and Xception on smaller, domain-specific datasets. This helps reduce training time and computational requirements while improving generalization [13]. Additionally, hybrid models that combine CNNs for spatial data with LSTMs for sequential analysis are being used to fuse multimodal sensor inputs and improve the robustness of defect prediction systems.

The next frontier in intelligent defect monitoring lies in real-time implementation through cyber-physical systems (CPS). With advancements in edge computing, it is now feasible to deploy ML/DL models directly on the machine for live inference. In such systems, anomaly detection triggers feedback loops that can autonomously adjust process parameters or terminate faulty builds, thus reducing material waste and production time [14]. Real-time monitoring also enables layer-wise quality assurance, improving traceability and part certification in industries with stringent safety standards.

Despite the substantial progress, several challenges remain. One major issue is the lack of large, labeled datasets for supervised learning, particularly for rare defect types. Most ML/DL models struggle to generalize across different machines, materials, and geometric configurations. Model interpretability is another concern, especially in high-risk domains like aerospace and medical implants, where black-box decisions may not be acceptable. Computational demands and the need for specialized hardware also pose barriers to real-time deployment on standard AM machines. To address these challenges, researchers are exploring avenues such as semi-supervised and unsupervised learning, federated learning, model compression, and physics-informed AI [15, 16].

The fusion of artificial intelligence and AM is transforming the way defects are monitored, analyzed, and mitigated. Data-driven models, particularly those based on ML and DL, are enabling a shift from reactive quality control to proactive and predictive defect prevention. This chapter delves into the application of these models in the context of thin-walled AM structures. It provides a comprehensive overview of the ML and DL techniques used for defect classification, anomaly detection, and real-time process monitoring. The discussion is enriched with real-world case studies, model architectures, and future research opportunities, offering a valuable resource for researchers, engineers, and practitioners aiming to advance intelligent manufacturing.

5.2 FEATURES MEASUREMENT IN THE ADDITIVELY MANUFACTURED COMPONENTS

The features of deposition in an AM process include (i) *single-layer single-track* deposition, (ii) *single-layer multi-track* deposition in which tracks are deposited with some overlap to cover the entire top surface of the substrate material, (iii) *multi-layer single-track* deposition in which layers are deposited successively to achieve the desired height, and (iv) *multi-layer multi-track* deposition in which layers are deposited along the width and height of substrate to achieve the desired width and height of complex geometries. Figure 5.1 depicts the dimensions of single-layer single-track and single-layer multi-track deposition geometries.

5.2.1 Manual Measurement of Deposition Geometry Features

In current practice, AM industries primarily rely on manual measurement methods to evaluate the deposition geometry of additively manufactured parts. Common approaches include the use of computer-aided design (CAD) modeling software and microscopy-based measurements. Cardoso et al. [17] manually measured the characteristics of the 40-layer wall deposited by the arc directed energy deposition process. They manufactured the wall of low-carbon steel (ER70S-6) and manually measured features such as actual and effective wall width (EWW) using a Zeiss microscope. Stinson et al. [18] manually measured the cross-sectional bead geometries obtained by metal inert gas (MIG) and cold metal transfer AM processes. They obtained the deposition geometry of the multi-layer wall by depositing ER70S-6 material and manually measured the actual and EWW using a Leica DMI8 microscope (Leica, Wetzlar, Germany), as shown in Figure 5.2.

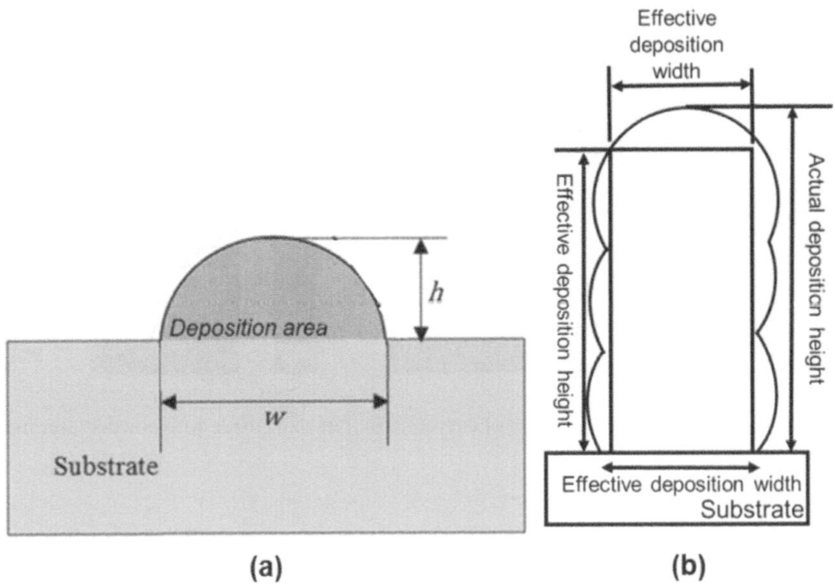

FIGURE 5.1 Features in deposition geometries: (a) single-layer single-track and (b) single-track multi-layer.

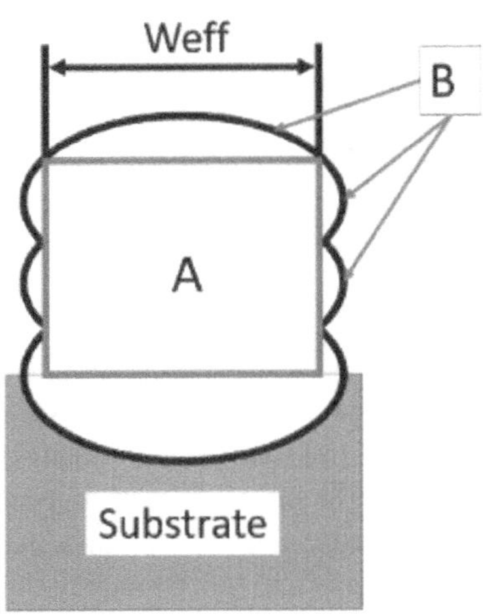

FIGURE 5.2 Geometry of multi-layer wall with effective wall area (A), actual wall area (B), and effective wall width (W_{eff}) [18].

Pereira et al. [19] used a manual measurement method to measure the thin-wall features manufactured by the laser metal deposition (LMD) process. They manufactured thin walls with different deposition materials such as SS316L and IN718 alloys. They measured the thickness and curvature of the wall in the depositions using optical microscopy, as shown in Figure 5.3.

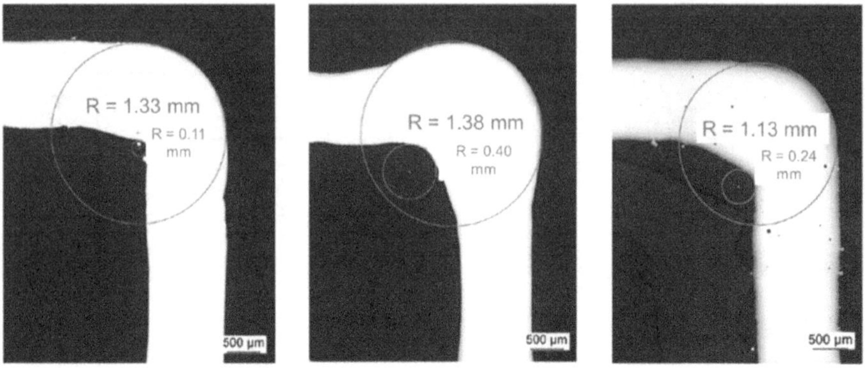

FIGURE 5.3 Manual measurement of curvature in thin walls made with powder laser metal deposition [19].

Jhavar et al. [20] developed a microplasma transferred arc (µ-PTA) wire deposition process for additive layer manufacturing using 300 µm AISI P20 tool steel wire. The process achieved high deposition efficiency (87%) and defect-free multi-layer walls, making it a cost-effective alternative for die and mold remanufacturing. Figure 5.4 illustrates a cross-sectional micrograph of the deposited wall, showing total wall width, EWW, and areas used for calculating deposition efficiency and surface waviness, thereby demonstrating the geometrical quality of the deposition.

Tapoglou and Clulow [21] measured the features of deposition geometry of single-layer/single-track depositions using an optical microscope, as shown in Figure 5.5. Using a similar method, Shim et al. [22] manually measured the features of single-layer single-track and single-track multi-layer deposition geometries. They found a high level of uncertainty in the manual measurement method.

5.2.2 Image Processing Approach for Feature Measurement

In AM, one of the important tasks is ensuring the dimensional accuracy of manufactured parts and components. The image processing method has established itself as a viable technique for measuring the dimensions of AM-manufactured components. Image processing methods have also found acceptance in other industrial applications, such as calibration of cameras and other imaging devices, characterization of microstructures, and measurement of features and dimensions of mechanical components. Detection and extraction of features from real images are difficult due to noise such as shadows, poor lighting during image acquisition, and irregularly shaped contours. The amount of information available in poor-quality images is not enough for feature extraction. Mechanical applications generally have irregular and complex geometries. Even though high-resolution equipment is used for capturing irregular and complex geometries, shape extraction is still a difficult task. These challenges necessitate the development of an image processing method capable of using these high-resolution images to extract irregular and complex features of mechanical components.

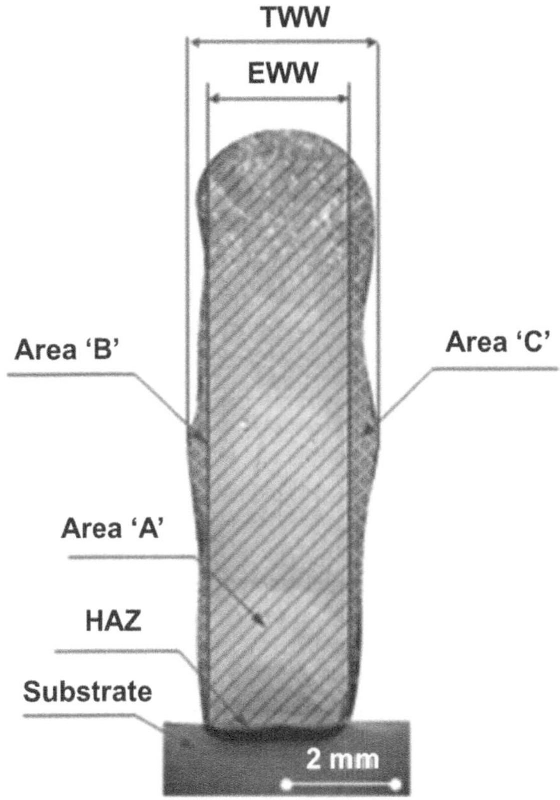

FIGURE 5.4 Optical micrograph showing various parameters used to judge the quality of multi-layer wall deposition by the μ-PTA process [20].

FIGURE 5.5 Manual measurement of single-layer deposition geometry using an optical microscope [21].

Past research suggests using image processing approaches like edge-following method, genetic algorithm method, and the Hough transform method for feature identification of mechanical items [23–25]. A feature recognition method for automatic detection of concave and convex shapes from tessellated CAD geometry was developed by Zhang and Li [26]. The process is carried out in two steps. Identifying local concave and convex shapes using the classification of triangular facets in an STL file is the first step, after which the Gaussian image formation and point distribution distance algorithm are used to detect shapes. The image processing method has been used to investigate defect formation and melt pool dynamics for the laser AM process by Leung et al. [27]. Real-time data for the deposition process has been captured using an image acquisition system connected with the machine used for laser AM. Mechanical attributes of coatings generated using the air plasma spraying process of ceria and yttria co-stabilized zirconia have been examined using Alfano et al. [28]. X-ray diffraction is used for analyzing the phases of coating material. For the examination of coating porosity and microstructure of coating material, a combination of SEM and image analysis has been used; on the other hand, in welding applications, past works use a combination of mechanical measuring instruments with image processing software. Automated online detection of defects in deposits created using the selective laser melting process has been done using an image processing approach by Caggiano et al. [29]. The process has been automated using a model based on a deep CNN applied to acquired images. Features of deposition geometry for the gas metal arc (GMA) welding process have been measured by Thao and Kim [30] using a microscope integrated with image processing software to obtain the height and width of deposition geometry. For the same GMA welding process, an optical microscope has been used for bed geometry feature measurement by Sathiya et al. [31]. The use of the Canny edge detection technique for analyzing deposition geometry obtained using a process based on robotic metal arc gas deposition has been developed by Cao et al. [32]. Image processing is used to obtain deposition geometry, which is then fitted with different curve functions like Gaussian, Logistic, parabola, and sine functions. An image processing method based on pixel intensity level is used for measuring deposition geometry dimensions by Singh et al. [33], which helped extract information from images.

The application of image processing methods for enhancing AM process performance is a major field explored by various researchers globally. The process of laser cladding-based AM used for building aircraft engine components using Inconel 718 by Tabernero et al. [34] captures images using a microscope on which image processing is applied for quantification of porosity, its size, and relative area. Putthawo et al. [35] applied the image processing method on laminated object manufacturing for tool path generation for adaptive cross-hatch. Image processing served as an information source to determine patterns that are usable for adaptive cross-hatching and intersection points while generating the tool path. Straub [36] used image analysis software to overcome challenges in defect detection for components manufactured using AM. In his work, the effectiveness of different cameras and image processing software has been utilized to evaluate the progress of defect detection occurring because of the poor quality of images obtained during the deposition process. The image processing method has been used to evaluate and control changes in grains' morphology by Wang et al. [37] for heavy-duty metallic components manufactured using the AM process based on LMD. Commercial image processing software, namely ImageJ and Photoshop CS, is used in their work. A novel image processing method for direct CAD slicer is developed by Vaidya and Anand [38] and is known as IPSlicer. Their developed method is useful for manufacturing complex geometry components using CAD data without intermittent conversion to STL files. The validation has been done using a snapshot of the product's image taken in the direction perpendicular to the movement of the heat source and along the section of the two-dimensional data for the bounding box. Tracing of contour points along the heat source movement by using the contour mapping algorithm identifies the performing boundary and its location. For components built using the AM process based on powder bed fusion, microstructure characterization and defect detection have been done using a multifractal method by Yao et al. [39]. Variations occurring in inherent images of components produced using AM are captured using this method, for which the resolution of densities and irregularities providing characterization data in multifractal spectrum form has been made. These spectrums denote image profile heterogeneity and are used to identify the shape and dimensions of defects of damaged products of AM. These past works establish the wide use and acceptance of image processing in different mechanical-based applications for process efficiency improvement, microstructure characterization, tolerance and dimensioning of geometry, and extraction of complex geometry features of mechanical components.

5.3 OPTIMIZATION OF THE ADDITIVE MANUFACTURING PROCESS

The importance of optimizing parameters of deposition geometry lies in the fact that it directly impacts the quantity and quality of depositions, which in turn enhances and improves the accuracy and productivity of AM components. Zhang and Zhou [40] optimized deposition process parameters for the magnetic field-assisted laser welding process. Using regression analysis, a relationship was developed between input welding process parameters and weld bead geometry characteristics. The obtained equation was used in the particle swarm optimization (PSO) algorithm to optimize weld bead geometry. Lestan et al. [41] obtained the optimized machine process parameters to improve the surface roughness value for laser-based additive manufacturing using genetic programming and non-dominated sorting methods. Yanxi et al. [42] investigated the weld bead appearance by predicting bead geometry parameters in terms of process parameters using a neural network. The model was further integrated with the genetic algorithm to optimize weld bead appearance. Dey et al. [43] generated a relationship between weld bead geometry parameters and electron beam melting process parameters using regression analysis. Geometry parameters were optimized using a regression model in the genetic algorithm. Kim et al. [44] used the genetic algorithm-based optimization method with a second-order regression equation for optimizing the Gas Metal Arc Welding (GMAW) process parameters. Lin [45] used the Taguchi method based on gray relational analysis and neural networks to optimize the GMAW process parameters. Nagaraju et al. [46] carried out optimization of process parameters for optimizing the depth of penetration and the heat-affected zone of the tungsten inert gas (TIG) welding process using response surface methodology with a genetic algorithm. Katherasan et al. [47] developed a neural network-based approach to predict the weld bead geometry parameters. They integrated this approach with the PSO algorithm for the flux-cored arc welding (FCAW) process.

Further, Senthilkumar et al. [48] used this approach in the genetic algorithm to optimize the welding process parameters of the FCAW process. Singh et al. [49] gave a fuzzy logic-based desirability function method to optimize bead deposition geometry obtained by submerged arc welding (SAW). In comparison, Da Silva et al. [50] generated a model to determine the dimension of weld bead overlap during the multi-track deposition process and to optimize SAW weld bead deposition process parameters. From a sustainability point of view, Vimal et al. [51] generated a neural network model to predict spatter losses, fume generation rate, energy consumption, and amount of slag using process parameters of the SAW process. The model was integrated with fuzzy gray relational analysis to optimize process parameters. Chaki et al. [52] predicted the strength of deposition in terms of process parameters of laser-MIG hybrid welding using the artificial neural network (ANN). The developed ANN model was combined with a genetic algorithm, simulated annealing, and Quasi-Newton line search methods to optimize process parameters. Orishich et al. [53] optimized the input energy level value to minimize internal defects for laser, microplasma, and laser-microplasma hybrid welding processes. Chandrasekhar et al. [54] used the genetic algorithm to optimize the penetration depth formed during the laser-TIG welding process. The multi-regression model in terms of process parameters was used in the algorithm for optimization. It can be concluded that optimization of deposition geometry in terms of deposition process parameters is essential because it defines the quality of the deposition. However, in past work, most models (i.e., Regression equation, ANN, etc.) used for optimization are experiment-based.

5.4 DEFECT DETECTION IN THE ADDITIVELY MANUFACTURED COMPONENTS

The structural integrity of AM-manufactured parts has been affected significantly by defects, AM process parameters, and the resultant mechanical properties of AM-manufactured parts. The operational performance of AM-manufactured components used in aerospace and biomedical industries has to be reliable even when subjected to a variety of dynamic and complex loadings under different environmental conditions. These loadings are often multiaxial, and the presence of defects within the additively manufactured parts can cause the failure of components under multiaxial loading conditions. The effect of defects under static loading is less significant; however, they greatly affect fatigue performance under cyclic loading. Metal AM parts consist of an array of defects such as LOF defects, gas porosities, cracks, and rough texture. These defects promote premature damage in the additively manufactured components. Therefore, to minimize or avoid the damage in metal AM, these defects have to be characterized, analyzed, and categorized. The following subsection explains the method used for characterizing, analyzing, and categorizing defects:

5.4.1 Image Processing

One approach used by the AM industries to detect defects is "Image Processing". This approach is mainly used for characterizing and analyzing the defects in the additively manufactured components. Image processing has wide applicability in the feature measurement of additively manufactured components. However, it can also

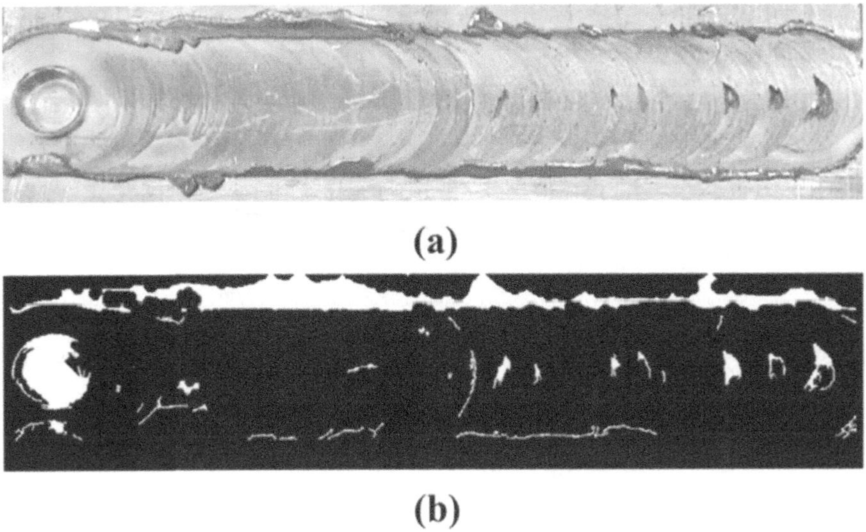

FIGURE 5.6 Image processing approach to detect defects in (a) friction stir welding and (b) defect in welding [55].

characterize and analyze defects in welding and AM applications. Ranjan et al. [55] developed an image processing approach to detect voids, grooves, cracks, keyholes, and flash in the friction steel welding application. Figure 5.6 shows the researchers' image processing approach to detect defects in friction stir welding. The developed approach aims to improve the weld quality using a monitoring system of welding parameters based on online feedback control.

Nacereddine et al. [56] developed a digital image processing approach to detect the arc welding-based deposition defects. Figure 5.7 depicts the methodology of digital image processing (Figure 5.7a) and detected defects (Figure 5.7b). The developed technique has been well developed so that improvement in both operating time and inspection process can be obtained. It attempted the development and implementation of algorithms of digital image processing based on both local and global approaches for automated detection of defects in arc welding-based depositions.

Tian et al. [57] developed a digital image processing method for automating the defect detection process in X-ray images of the depositions done in welding applications. The automatic defect detection image processing approach, which uses image data fusion, incorporating different techniques like weld district extraction, segmentation with dynamic threshold, wave profile analysis, and edge extraction, has been determined that defects inducing a sudden change over a predefined extent of the image intensity can be segmented irrespective of their size, shape, location, and number.

Nacereddine et al. [58] used X-ray images to construct a weld defect descriptor set and to recognize them using the image processing approach for the depositions made in an arc welding process. Figure 5.8 depicts the detection of defects such as lack of penetration and aligned inclusions. Geometric invariant moments form this

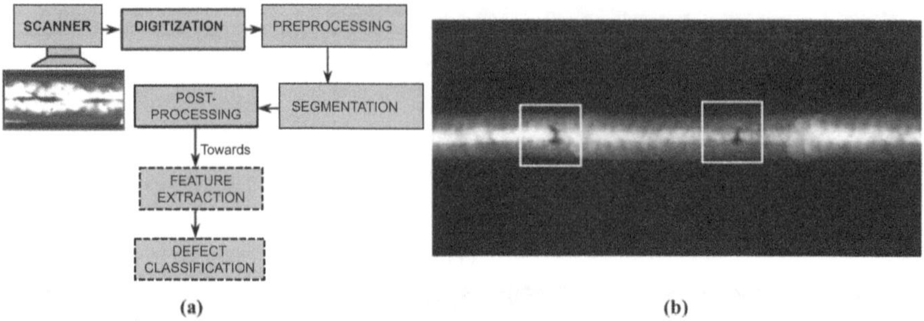

FIGURE 5.7 Image processing methodology to detect defects: (a) methodology and (b) detected defect in welding [56].

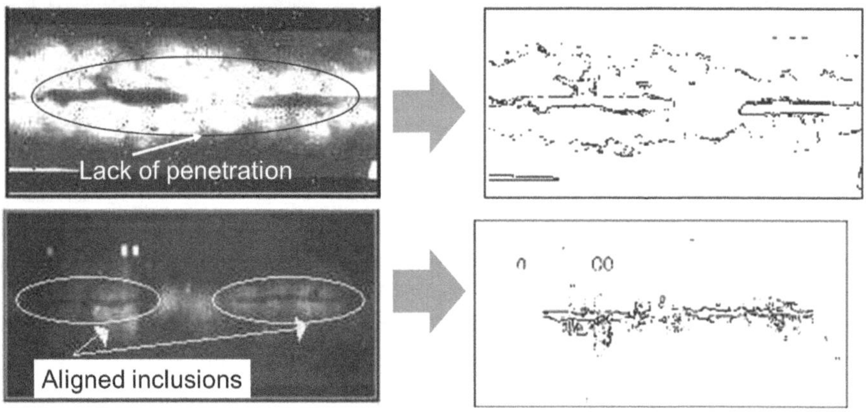

FIGURE 5.8 Image processing approach to detect defects such as penetration and aligned inclusions in an arc welding deposition process [58].

approach's basis; these are also insensitive to common geometric transformations. Once the computation of geometric invariant features is done, an image processing approach is used to detect image defects in a planar or volumetric format.

Pavan et al. [59] used image processing methodology on the image captured by X-ray CT. It was used to capture porosity content, pore count, and volume distributions within the components manufactured by laser sintering for polymers. Figure 5.9 depicts the image processing approach on the CT scan captured images to quantify the porosity within the additively manufactured components. Foster et al. [60] used the X-ray CT scan method to capture the images of internal defects in the powder bed fusion additively manufactured components. They used the image processing method to detect the internal defects manually. This developed technique is based on collecting and analyzing images of fused and pre-placed powder layers taken under oblique illumination.

Jacobsmuhlen et al. [61] developed a high-resolution imaging methodology used to inspect defects in the laser beam melting manufactured components. They used image processing to detect the edges' outer region elevation in the additively

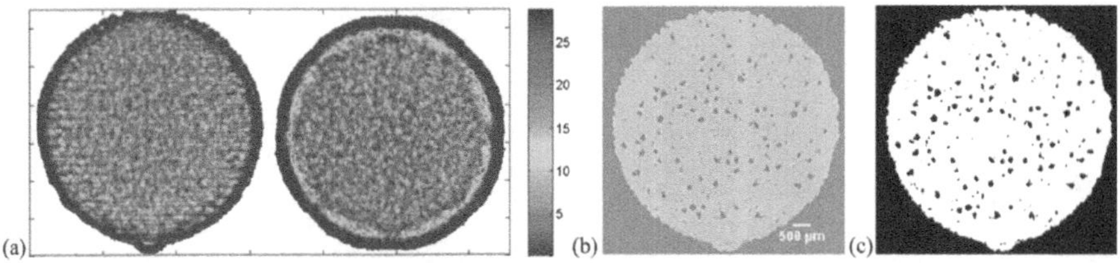

FIGURE 5.9 Image processing to detect (a) porosity volume, (b) and (c) porosity count and content, respectively [59].

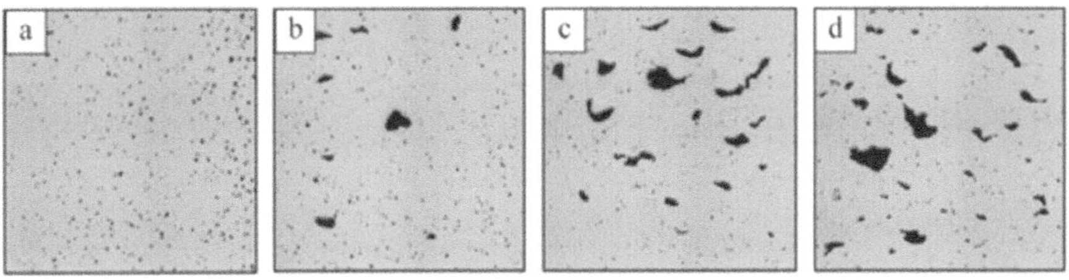

FIGURE 5.10 Microscopic image processing approach detecting defects such as (a–b) metallurgical pores and (c–d) keyhole porosity in the selective laser melting additive manufacturing process [63].

manufactured components. Evaluation of the resolving power of the developed imaging system is done. It has been found that it is capable of inspecting the results of the process on a microscopic scale. It was also found that topological flaws can be detected using their system, and inspection of the surface quality of built layers can be done. Similarly, Kleszczynski et al. [62] used a high-resolution imaging method to detect irregularities in the laser beam melting additively manufactured components.

Aboulkhair et al. [63] developed a microscopic image processing method to detect and quantify defects in the selective laser melting additively manufactured components. They used this approach to detect metallurgical pores and keyhole porosity in the components manufactured using AM, as shown in Figure 5.10.

5.4.2 Machine Learning

Apart from image processing, data-driven methods such as ML have also given certain advantages, and high accuracy in detecting defects in the additive manufactured components [64]. Mohammadi et al. [65] compared different ML techniques and analyzed their performance for online detection of defects in the process of laser powder bed fusion (L-PBF). Figure 5.11 shows the various classifications in the components manufactured using the L-PBF process. It also depicts the defects such as cracks and porosities, as shown in Figure 5.11(b)–(c). They used the hierarchical K-means method, as shown in Figure 5.12, to classify and categorize the defects in additively manufactured components.

FIGURE 5.11 Samples with (a) no defects (class 1), (b) cracks only (class 2), and (c) both cracks and porosities (class 3) [65].

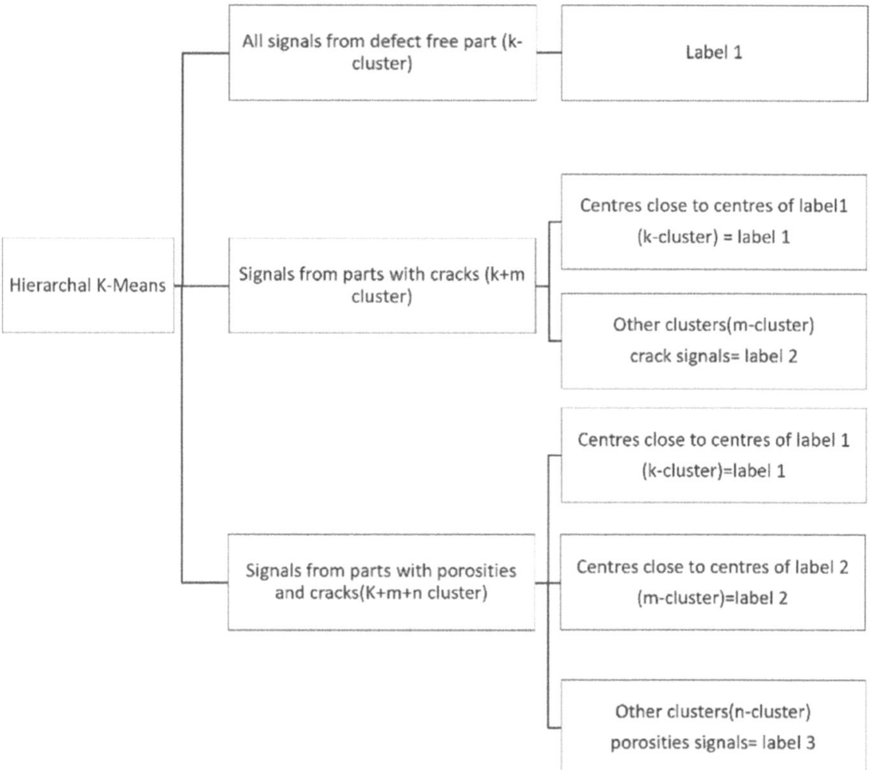

FIGURE 5.12 Hierarchical K-means method used for defect classification [65].

Scime and Beuth [66] developed a trained computer vision algorithm for detecting and classifying anomalies in the L-PBF additive manufactured components. Figure 5.13 shows the confusion matrix with the contribution of each type of ground-truth anomaly and anomaly guesses (in percentage) made using the algorithm for the L-PBF process. Their result showed that the detection of anomalies and the implemented classification using an unsupervised ML algorithm could function

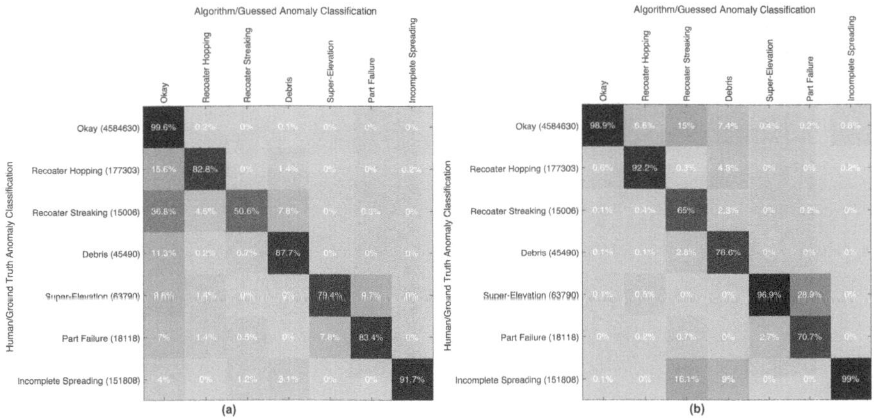

FIGURE 5.13 A confusion matrix showing (a) the percentage of each ground-truth anomaly and (b) the percentage of anomaly guesses made by the algorithm for the L-PBF process [66].

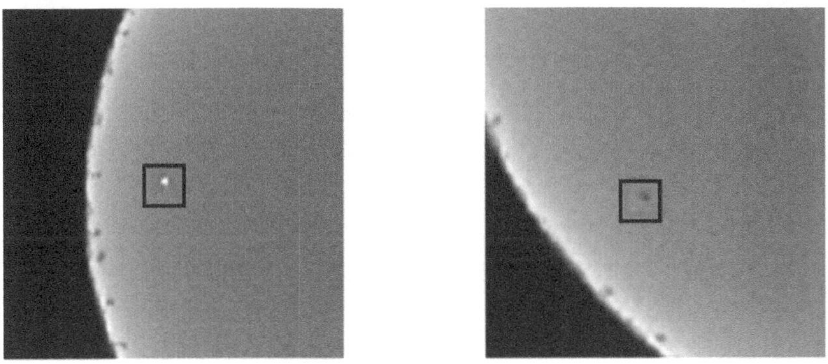

FIGURE 5.14 Voids or pores detected by lower-intensity voxels (left) and higher-intensity voxels (right) [67].

using a moderately sized image patch training database. Gobert et al. [67] applied supervised ML to detect high-resolution imaging defects during the metallic powder bed fusion process. This approach detected small pores or voids in the additive manufactured components, as shown in Figure 5.14. They found that, after proper classifier training, during cross-validation experiments, an accuracy of more than 80% for in situ detection of defects was demonstrated.

Okaro et al. [68] introduced an algorithm of ML for automated fault detection in products made using AM. A semi-supervised approach is developed, which can use data from both builds during training, i.e., build with certified resulting components and builds with unknown resulting component quality. This helped improve the cost efficiency of this approach, specifically in scenarios where part certification is time-consuming and costly. Their study particularly analyzed builds from L-PBF. Westphal and Seitz [69] developed a complex TL method enabling the automated

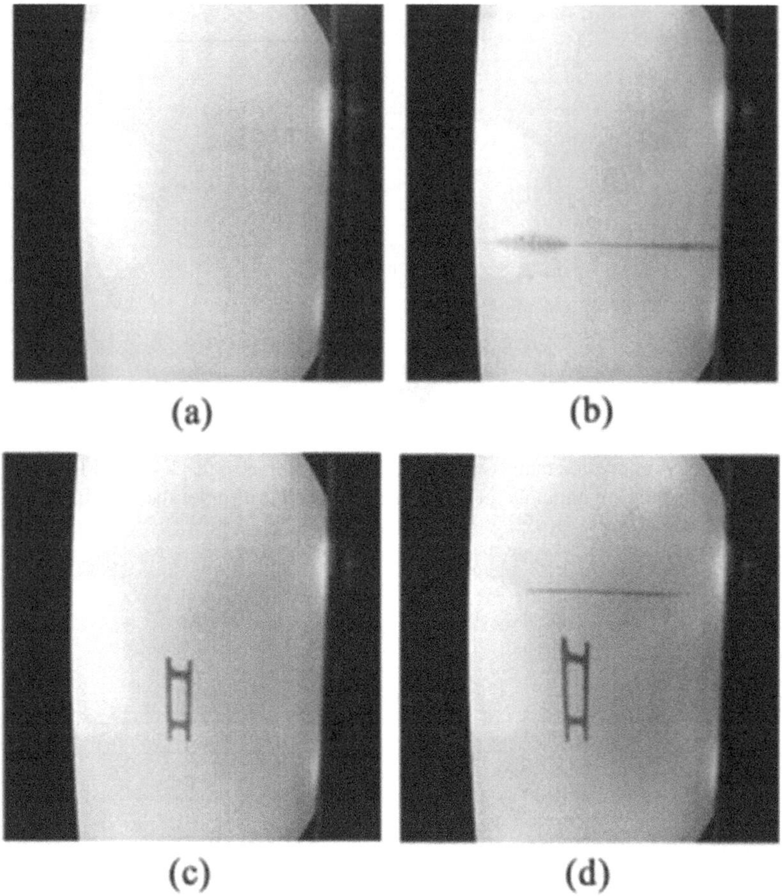

FIGURE 5.15 Dataset of (a) a powder bed without defects, (b) a powder bed with irregularities, (c) a powder bed with a sintered element, and (d) a powder bed with a sintered element and irregularities [69].

classification of defects of the powder bed in the selective laser sintering process having small datasets as shown in Figure 5.15. They used the VGG16 and the Xception CNN model. In this, for the initialization, pre-trained weights were taken from the ImageNet dataset, and for classification of defective and good images, data obtained using the manufacturing of a part with an adapted classifier is used, as represented in Figure 5.16. These results highlighted that the Xception CNN-based defect detection offers a viable alternative for the most accurate prediction.

Xiao et al. [70] have developed a method to be used in the selective laser sintering process to detect three types of powder bed defects based on CNN. A digital camera is used to record the images of the powder bed, which were later analyzed using a dual-stage CNN. Identification of defect location was made in the first step, while in the second step, corresponding to each error location, error masks were generated. The use of this dual-stage CNN model helped improve error detection accuracy over

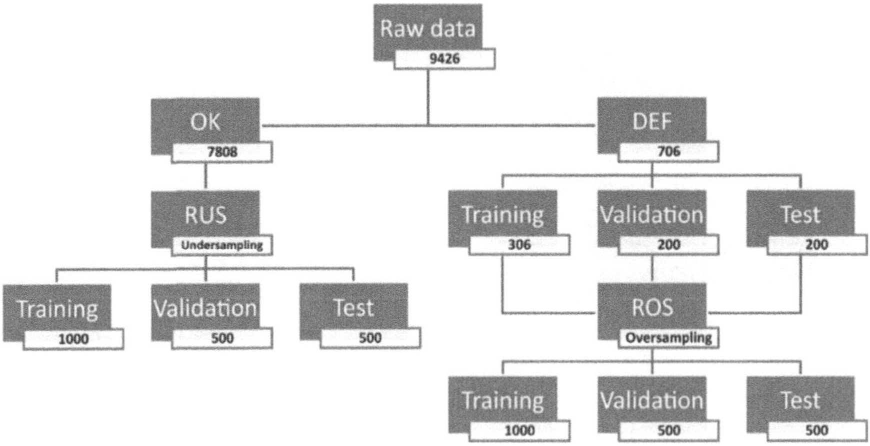

FIGURE 5.16 General procedure for creating the data structure described from the powder bed dataset [69].

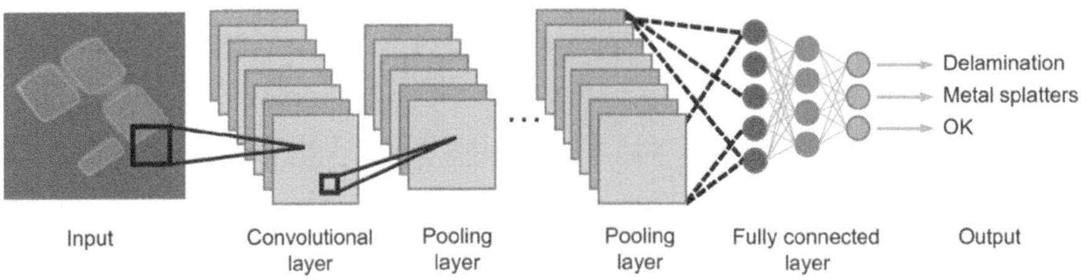

FIGURE 5.17 CNN architecture consisting of multiple convolutional layer blocks to detect defects such as delamination and metal splatters [71].

other methods. For simulating special error cases, pre-configured datasets have been used. However, details about the accuracy of dual-stage CNN on datasets of real manufacturing processes are not provided.

Baumgartl et al. [71] implemented a combination of off-axis in situ imaging and thermographic imaging in an L-PBF system. These images act as data sources for a CNN based on DL to identify printing errors. This developed model is suitable for thermographic in situ detection of defects for L-PBF processes, as represented in Figure 5.17, and reported 96% or higher accuracy levels. However, this CNN model could detect delamination and spatter types of defects occurring in the L-PBF process. Investigating other defects like unmelted powder, pores, cracks, and additional phenomena such as Selective Laser Sintering (SLS) was not done.

Ye et al. [72] developed a defect detection method for selective laser melting parts using a deep belief network framework. A simple classification structure does not have any feature extraction module or signal preprocessing module for its implementation. Experiments showed that acoustic signals could be utilized for quality monitoring, and a high defect detection rate among five melted states can be achieved using the Dynamic Bayesian Network (DBN) approach without signal preprocessing.

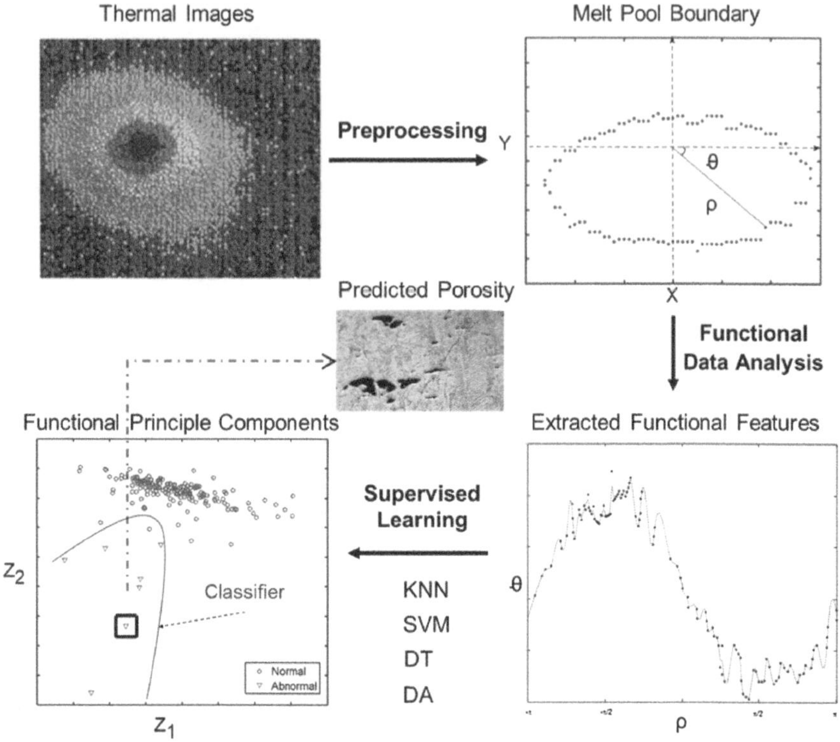

FIGURE 5.18 Porosity prediction procedures using supervised machine learning.

Khanzadeh et al. [73] utilized melt pool monitoring images to compare different supervised ML techniques as represented in Figure 5.18. The K-nearest neighbor algorithm was tested as the best and exhibited an accuracy of around 98% for real-time detection of melt pool anomaly and potentially microstructure anomaly.

5.5 MONITORING AND CONTROLLING OF THE ADDITIVE MANUFACTURING PROCESS

In situ monitoring and controlling of the deposition process in AM is a challenging and essential process. Simultaneously, there are challenges in characterizing melt pool shape and size [74]. These challenges in AM can be accomplished by CPS, gaining popularity with the onset of Industry 4.0. Image processing and ML have made this challenge achievable, and consequently, researchers are using it for feature measurement and defect detection in additively manufactured components. Some of them have used it for quality assessment of components, energy management, optimization of process parameters, and modeling complex scenarios [75].

Yetis and Karakose [76] developed an anomaly detection approach based on image processing for synchronous movement in CPS. They developed this approach for the movement of robot arms used in AM applications, as their work is divided into two sections: the implementation of an industrial robot system controlled remotely is the first step, while detection of unwanted movements and system control is the second step.

Shevchik et al. [77] had explored the use of acoustic emission (AE) for monitoring quality and used ML combined with a sensitive AE sensor, as shown in Figure 5.19. According to the process, classification based on the quality of manufactured components using the laser powder bed AM process as poor, medium, and high has been done. Training of the classifier was done using Spectral CNN for differentiating dissimilar acoustic features. Classification confidence obtained ranges between 83 and 89%.

Chen et al. [78] developed a method for rapid surface defect identification for AM components by combining ML and in situ point cloud processing. The development of the software platform has been carried out in-house with a multi-nodal architecture. Simultaneously, cloud-based processing involving different activities like filtering, segmentation, surface-to-point distance calculation, point clustering, and ML-based feature extraction was carried out. Experimental validation of this surface defect identification method was carried out, and an accuracy of 93.15% was obtained. Gaja and Liou [79] developed a novel defect monitoring system in LMD for real-time detection and classification of defects with the help of an AE sensor combined with unsupervised pattern recognition analysis (K-means clustering) along with principal component analysis (PCA). This technology can efficiently distinguish signals representing defects like cracks and porosities. The AE sensor was successfully used to detect cracks and pores occurring during LMD accurately. Grasso et al. [80] used a machine vision system in the visible range to develop a technique that detects and spatially identifies defects during the layer-wise process. Using a statistical descriptor that works on applying PCA on image data, identification of defective layer areas was made. Automatic defect detection had been achieved using

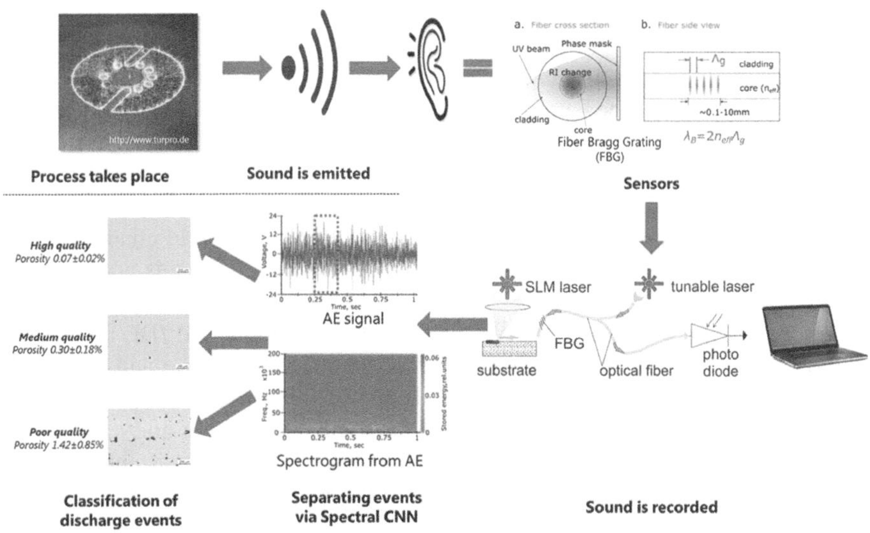

FIGURE 5.19 Procedure for proper quality monitoring in additive manufacturing [77].

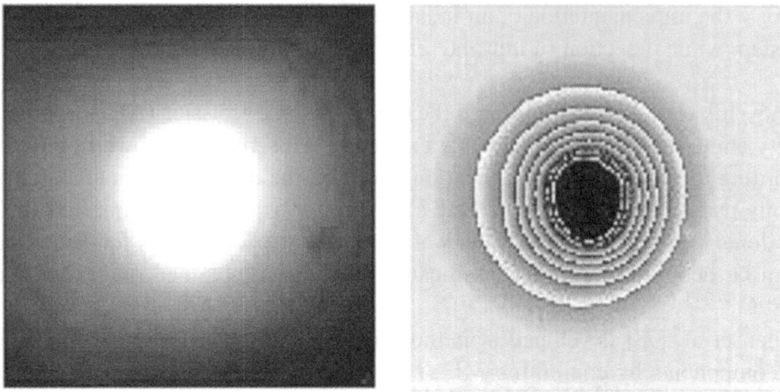

FIGURE 5.20 Infrared image acquired: (a) original infrared image and (b) gray level isotherms [81].

K-means clustering analysis. Hu and Kovacevic [81] developed a system working in a closed loop for controlling heat input and molten pool size for the laser-based additive manufacturing process based on an infrared image sensing approach. A coaxial camera is installed to capture infrared images, which enables observation to be made from the top of the molten pool, preventing any blocking and acquiring a full view field of the molten pool as represented in Figure 5.20(a). The temperature distribution in the molten pool and its surroundings is indicated by infrared images, specifically their gray level isotherms represented in Figure 5.20(b).

5.6 CHALLENGES AND FUTURE SCOPE

5.6.1 Challenges

The integration of data-driven ML and DL models into AM processes for defect detection and feature measurement faces several notable challenges. One of the foremost issues is the scarcity of high-quality, annotated datasets, particularly for rare defects or complex geometries. The collection and labeling of such data are time-consuming and expensive. Moreover, AM processes exhibit significant variability due to differences in materials, machines, build conditions, and geometries, which hampers the generalizability of trained models across different setups.

Another major concern is the real-time implementation of ML/DL models. DL algorithms such as CNNs and RNNs are computationally intensive and may not be suitable for deployment on resource-limited edge devices used for in situ monitoring. Additionally, sensor data in industrial AM environments is often noisy and incomplete, making reliable data acquisition and preprocessing a persistent challenge. Furthermore, the absence of standardized protocols for data acquisition, model training, and performance evaluation makes it difficult to benchmark models across different research and industrial scenarios.

5.6.2 Future Scope

Despite these challenges, the future of data-driven approaches in AM is promising and full of opportunities. One significant direction is the development of multimodal learning models that integrate visual, thermal, acoustic, and force data to improve defect detection accuracy and robustness. Physics-informed ML presents another avenue, where domain-specific knowledge is embedded within data-driven models to enhance interpretability and reliability.

Moreover, TL and federated learning can enable the use of pre-trained models across different AM platforms, reducing the need for large labeled datasets and enhancing model adaptability. Self-supervised and semi-supervised learning techniques also hold great potential for extracting valuable insights from unlabeled data, further alleviating the burden of manual annotation.

With advancements in Edge AI, lightweight and efficient ML/DL models can be deployed directly on embedded hardware for real-time process control. The integration of AI models with digital twin systems will facilitate predictive maintenance, anomaly detection, and closed-loop control in smart manufacturing environments. Standardization of datasets and benchmarking practices will also be critical to accelerate innovation and adoption of intelligent defect monitoring systems in the AM industry.

5.7 SUMMARY

This chapter has presented an in-depth review of data-driven models for feature measurement and defect monitoring in AM, with a particular emphasis on their application to thin-walled and complex geometry structures. The discussion began with an exploration of conventional manual and image-based feature measurement methods, highlighting their limitations in terms of accuracy, scalability, and automation. Image processing techniques have shown promise in automating the extraction of geometrical features and detecting irregularities during and after the build process, but their performance is often constrained by the quality of acquired data and the complexity of deposition geometries.

With the emergence of Industry 4.0 and the proliferation of real-time data acquisition technologies, there is a significant shift toward leveraging ML and DL for intelligent defect detection. Supervised learning methods such as decision trees, SVM, and neural networks have enabled the classification of known defect types, while unsupervised approaches including clustering and autoencoders have proven valuable in discovering anomalous patterns without requiring extensive labeled datasets. Advanced DL models like CNNs and RNNs have demonstrated state-of-the-art performance in analyzing spatial and temporal data, making them highly suitable for in situ monitoring of melt pool behavior, thermal gradients, and structural evolution.

Furthermore, the chapter reviewed several studies demonstrating the integration of data-driven models with optimization algorithms, control systems, and cyber-physical infrastructure, thereby enabling real-time monitoring and closed-loop feedback during AM processes. The use of AEs, thermal imaging, X-ray CT, and machine vision has enabled the generation of large datasets that can be harnessed to train robust and predictive AI models.

However, the successful deployment of these technologies is still met with various challenges such as data scarcity, model generalization, computational constraints, and standardization issues. The chapter identified future research directions, including multimodal data fusion, physics-informed learning, federated learning, and Edge AI, all of which are expected to play a pivotal role in achieving autonomous, high-precision, and defect-free AM.

In conclusion, the adoption of data-driven ML/DL approaches in AM represents a transformative shift from reactive quality inspection to predictive and adaptive manufacturing. By harnessing these intelligent models, the industry is poised to achieve improved part quality, enhanced process stability, and increased production efficiency—hallmarks of a smart, connected, and sustainable manufacturing ecosystem.

REFERENCES

1. Gibson, I., Rosen, D. W., and Stucker, B. *Additive Manufacturing Technologies*, 2nd Edition, Springer Science+Business Media New York, 2015.
2. Frazier, W. E. Metal additive manufacturing: A review, *Journal of Materials Engineering and Performance*, 23(6), 1917–1928, 2014.
3. Tapia, G., and Elwany, A. A review on process monitoring and control in metal-based additive manufacturing, Journal of Manufacturing Science and Engineering, 136(6), 060801, 2014.
4. Grasso, M., and Colosimo, B. M. Process defects and in-situ monitoring methods in metal powder bed fusion: A review, *Measurement Science and Technology*, 28(4), 044005, 2017.
5. Scime, L., and Beuth, J. Anomaly detection and classification in a laser powder bed additive manufacturing process using a trained computer vision algorithm, *Additive Manufacturing*, 19, 114–126, 2018.
6. Mohammadi, M., et al. Machine learning approaches for defect detection in laser powder bed fusion, *Journal of Manufacturing Processes*, 67, 333–347, 2021.
7. Goodfellow, I., Bengio, Y., and Courville, A. *Deep Learning*, MIT Press, 2016.
8. Ye, R., et al. Real-time monitoring and anomaly detection in additive manufacturing using acoustic signals and deep belief networks, *Additive Manufacturing*, 35, 101307, 2020.
9. Xiao, G., et al. A deep learning-based method for real-time defect detection in SLS powder beds, *IEEE Transactions on Industrial Electronics*, 67(7), 5981–5991, 2020.
10. Westphal, T., and Seitz, H. Transfer learning for defect classification in powder bed fusion, *Procedia CIRP*, 93, 226–231, 2020.
11. Baumgartl, H., et al. In-situ defect detection using thermographic imaging and CNNs in LPBF, *Materials*, 14 (5), 1029, 2021.
12. Yao, S., et al. Multifractal image processing for microstructural defect detection in laser powder bed fusion, *Applied Sciences*, 10(1), 181, 2020.
13. Westphal, T., et al. Using transfer learning for defect detection in SLM using pre-trained CNNs, *Procedia Manufacturing*, 47, 1118–1123, 2020.
14. Grasso, M., and Colosimo, B. M. Closed-loop control for additive manufacturing: Challenges and opportunities, *Journal of Manufacturing Systems*, 48, 69–76, 2018.
15. Qin, J., Liu, Y., and Grosvenor, R. A categorical framework of manufacturing for Industry 4.0 and beyond, *Procedia CIRP*, 52, 173–178, 2016.
16. Tapia, G., et al. Data-driven modeling of process parameters and defects in metal AM, *Additive Manufacturing*, 13, 143–157, 2017.

17. Cardoso, A., Assunção, E., and Pires, I. Study of a hardfacing flux-cored wire for arc directed energy deposition applications, *International Journal of Advanced Manufacturing Technology*, 118, 3431–3442, 2021. https://doi.org/10.1007/s00170-021-08144-6.
18. Stinson, H., Ward, R., Quinn, J., and McGarrigle, C. Comparison of properties and bead geometry in mig and cmt single layer samples for waam applications. *Metals*, 11(10), 1530. https://doi.org/10.3390/met11101530.
19. Pereira, J.C., Borovkov, H., Zubiri, F., Guerra, M.C., and Caminos, J. Optimization of thin walls with sharp corners in SS316L and IN718 alloys manufactured with laser metal deposition. *Journal of Manufacturing and Materials Processing*, 5(1), 5. https://doi.org/10.3390/jmmp5010005.
20. Jhavar, S., Jain, N.K., and Paul, C.P. Development of micro-plasma transferred arc (-PTA) wire deposition process for additive layer manufacturing applications. *Journal of Materials Processing Technology*, 214, 1102–1110, 2014. https://doi.org/10.1115/1.4040324
21. Tapoglou, N., and Clulow, J. Investigation of hybrid manufacturing of stainless steel 316L components using direct energy deposition. *Proceedings of the Institution of Mechanical Engineers, Part B: Journal of Engineering Manufacture*, 235, 1633–1643, 2021. https://doi.org/10.1177/0954405420949360.
22. Shim, D.S., Baek, G.Y., Seo, J.S., Shin, G.Y., Kim, K.P., and Lee, K.Y. Effect of layer thickness setting on deposition characteristics in direct energy deposition (DED) process. *Optics & Laser Technology*, 86, 69–78, 2016. https://doi.org/10.1016/j.optlastec.2016.07.001.
23. McLaughlin, R.A. Randomized hough transform: Improved ellipse detection with comparison, *Pattern Recognition Letters*, 19, 299–305, 1998. https://doi.org/ 10.1016/S0167-8655(98)00010-5.
24. Kiryati, N., Eldar, Y., and Bruckstein, A.M. A probabilistic hough transform. *Pattern Recognition*, 24, 303–316, 1991. https://doi.org/10.1016/0031-3203(91)90073-E.
25. Duda, R.O., and Hart, P.E. Use of the hough transformation to detect lines and curves in pictures. *Communications of the ACM*, 15, 11–15, 1972. https://doi.org/10.1145/361237.361242.
26. Zhang, J. and Li, Y. Region segmentation and shape characterisation for tessellated CAD models. *International Journal of Computer Integrated Manufacturing*, 29, 907–915, 2016. https://doi.org/10.1080/0951192X.2015.1130249.
27. Leung, C.L.A., Marussi, S., Towrie, M., Atwood, R.C., Withers, P.J., Lee, P.D., The effect of powder oxidation on defect formation in laser additive manufacturing. *Acta Materialia*, 166, 294–305, 2019. https://doi.org/10.1016/j.actamat.2018.12.027.
28. Alfano, M., Di Girolamo, G., Pagnotta, L., and Sun, D. Processing, microstructure and mechanical properties of air plasma-sprayed ceria-yttria co-stabilized zirconia coatings. *Strain*, 46, 409–418, 2010. https://doi.org/10.1111/j.1475-1305.2009.00659.x.
29. Caggiano, A., Zhang, J., Alfieri, V., Caiazzo, F., Gao, R., and Teti, R. Machine learning-based image processing for on-line defect recognition in additive manufacturing. *CIRP Annals*, 68, 451–454, 2019. https://doi.org/10.1016/j.cirp.2019.03.021.
30. Thao, D.T., and Kim, I.S. Interaction model for predicting bead geometry for Lab Joint in GMA welding process. *Archives of Computational Materials Science and Surface Engineering*, 1, 237–244, 2009.
31. Sathiya, P., Aravindan, S., Ajith, P., Arivazhagan, B., and Haq, A.N. Microstructural characteristics on bead on plate welding of AISI 904 L super austenitic stainless steel using Gas metal arc welding process. *International Journal of Engineering, Science and Technology*, 2(6), 189–199, 2010. https://doi.org/10.4314/ijest.v2i6.63710.

32. Cao, Y., Zhu, S., Liang, X., and Wang, W. Overlapping model of beads and curve fitting of bead section for rapid manufacturing by robotic MAG welding process. *Robotics and Computer-Integrated Manufacturing*, 27, 641–645, 2011. https://doi.org/10.1016/j.rcim.2010.11.002.
33. Singh, A.K., Dey, V., Rai, R.N., and Debnath, T. Weld bead geometry dimensions measurement based on pixel intensity by image analysis techniques. *Journal of The Institution of Engineers (India): Series C*, 100, 379–384, 2019. https://doi.org/10.1007/s40032-017-0392-8.
34. Tabernero, I., Lamikiz, A., Martínez, S., Ukar, E., and Figueras, J., Evaluation of the mechanical properties of Inconel 718 components built by laser cladding. *International Journal of Machine Tools and Manufacture*, 51, 465–470, 2011. https://doi.org/10.1016/j.ijmachtools.2011.02.003.
35. Putthawong, S., Koomsap, P., and Chansri, N. Image-based adaptive crosshatch toolpath generation for laminated object manufacturing, *Virtual and Physical Prototyping*, 9(4), 233–249, 2014. https://doi.org/10.1080/17452759.2014.954823.
36. Straub, J. Initial work on the characterization of additive manufacturing (3D printing) using software image analysis, *Machines*, 3, 55–71, 2015. https://doi.org/10.3390/machines3020055.
37. Wang, T., Zhu, Y.Y., Zhang, S.Q., Tang, H.B., and Wang, H.M., Grain morphology evolution behavior of titanium alloy components during laser melting deposition additive manufacturing. *Journal of Alloys and Compounds*, 632, 505–513, 2015. https://doi.org/10.1016/j.jallcom.2015.01.256.
38. Vaidya, R., and Anand, S. Image processing assisted tools for pre- and post-processing operations in additive manufacturing. *Procedia Manufacturing*, 5, 958–973, 2016. https://doi.org/10.1016/j.promfg.2016.08.084.
39. Yao, B., Imani, F., Sakpal, A.S., Reutzel, E.W., and Yang, H. Multifractal analysis of image profiles for the characterization and detection of defects in additive manufacturing. *Journal of Manufacturing Science and Engineering*, 140, 031014, 2018. https://doi.org/10.1115/1.4037891.
40. Zhang, F. and Zhou, T. Process parameter optimization for laser-magnetic welding based on a sample-sorted support vector regression, *Journal of Intelligent Manufacturing*, 30, 2217–2230, 2019. https://doi.org/10.1007/s10845-017-1378-3.
41. Lestan, Z., Klancnik, S., Balic, J., and Brezocnik, M. Modeling and design of experiments of laser cladding process by genetic programming and nondominated sorting. *Materials and Manufacturing Processes*, 30, 458–463, 2015. https://doi.org/10.1080/10426914.2014.973586.
42. Yanxi, Z., Xiangdong, G. and Seiji, K. Weld appearance prediction with BP neural network improved by genetic algorithm during disk laser welding. *Journal of Manufacturing Systems*, 34, 53–59, 2015. https://doi.org/10.1016/j.jmsy.2014.10.005.
43. Dey, V., Pratihar, D.K., Datta, G.L., Jha, M.N., Saha, T.K., and Bapat, A.V. Optimization and prediction of weldment profile in bead-on-plate welding of Al-1100 plates using electron beam. *International Journal of Advanced Manufacturing Technology*, 48, 513–528, 2010. https://doi.org/10.1007/s00170-009-2307-1.
44. Kim, D., and Rhee, S. Optimization of arc welding process parameters using a genetic algorithm, *Welding Journal*, 80(7), 184–189, 2001.
45. Lin, H.L. The use of the Taguchi method with grey relational analysis and a neural network to optimize a novel GMA welding process. *Journal of Intelligent Manufacturing*, 23, 1671–1680, 2012. https://doi.org/10.1007/s10845-010-0468-2.
46. Nagaraju, S., Vasantharaja, P., Chandrasekhar, N., Vasudevan, M., and Jayakumar, T. Optimization of welding process parameters for 9Cr-1Mo steel using RSM and GA. *Materials and Manufacturing Processes*, 31, 319–327, 2016. https://doi.org/10.1080/10426914.2015.1025974.

47. Katherasan, D., Elias, J. V., Sathiya, P., and Haq, A.N. Simulation and parameter optimization of flux cored arc welding using artificial neural network and particle swarm optimization algorithm. *Journal of Intelligent Manufacturing*, 25, 67–76, 2014. https://doi.org/10.1007/s10845-012-0675-0.
48. Senthilkumar, B., Kannan, T., and Madesh, R. Optimization of flux-cored arc welding process parameters by using genetic algorithm. *International Journal of Advanced Manufacturing Technology*, 93, 35–41, 2017. https://doi.org/10.1007/s00170-015-7636-7.
49. Singh, A., Datta, S., Mahapatra, S.S., Singha, T., and Majumdar, G. Optimization of bead geometry of submerged arc weld using fuzzy based desirability function approach. *Journal of Intelligent Manufacturing*, 24, 35–44, 2013. https://doi.org/10.1007/s10845-011-0535-3.
50. Da Silva, M.M., Batista, V.R., Maciel, T.M., dos Santos, M.A., Brasileiro, T.L. Optimization of submerged arc welding process parameters for overlay welding. *Welding International*, 32, 122–129, 2018. https://doi.org/10.1080/09507116.2017.1347325.
51. Vimal, K.E.K, Vinodh, S., and Raja, A. Optimization of process parameters of SMAW process using NN-FGRA from the sustainability view point. *Journal of Intelligent Manufacturing*, 28, 1459–1480, 2017. https://doi.org/10.1007/s10845-015-1061-5.
52. Chaki, S., Shanmugarajan, B., Ghosal, S., and Padmanabham, G. Application of integrated soft computing techniques for optimisation of hybrid CO_2 laser-MIG welding process. *Applied Soft Computing*, 30, 365–374, 2015. https://doi.org/10.1016/j.asoc.2015.01.045.
53. Orishich, A.M., Malikov, A.G., Shelyagin, V.D., Khaskin, V.Y., and Chayka, A.A. Optimisation of the processes of laser, microplasma and hybrid laser–microplasma welding of aluminium alloys. *Welding International*, 30, 957–961, 2016. https://doi.org/10.1080/09507116.2016.1157338.
54. Chandrasekhar, N., Ragavendran, M., Ravikumar, R., Vasudevan, M., and Murugan, S. Optimization of hybrid laser–TIG welding of 316LN stainless steel using genetic algorithm. *Materials and Manufacturing Processes*, 32, 1094–1100, 2017. https://doi.org/10.1080/10426914.2017.1317793.
55. Ranjan, R., Khan, A.R., Parikh, C., Jain, R., Mahto, R.P., Pal, S., and Chakravarty, D. Classification and identification of surface defects in friction stir welding: An image processing approach. *Journal of Manufacturing Processes*, 22, 237–253, 2016. https://doi.org/10.1016/j.jmapro.2016.03.009.
56. Nacereddine, N. Weld defect detection in industrial radiography based digital image processing, *Presented at 3rd International Conference: Sciences of Electronic, Technologies of Information and Telecommunications*, March 27–31, 2005 – Tunisia.
57. Tian, Y., Du, D., Cai, G., Wang, L., and Zhang, H. Automatic defect detection in X-ray images using image data fusion. *Tsinghua Science & Technology*, 11, 720–724, 2006. https://doi.org/10.1016/S1007-0214(06)70255-3.
58. Nacereddine, N., Drai, R., and Benchaala, A. Weld defect extraction and identification in radiograms based neural networks. *Presented in 15th World Conference on Nondestructive Testing*, October 2000 – Roma, Italy.
59. Pavan, M., Craeghs, T., Verhelst, R., Ducatteeuw, O., Kruth, J.P., and Dewulf, W. CT-based quality control of Laser Sintering of Polymers. *Case Studies in Nondestructive Testing and Evaluation*, 6, 62–68, 2016. https://doi.org/10.1016/j.csndt.2016.04.004.
60. Foster, B.K., Reutzel, E.W., Nassar, A.R., Hall, B.T., Brown, S.W., and Dickman, C.J. *Optical, layerwise monitoring of powder bed fusion*, 2015, pp. 295–307.
61. Jacobsmuhlen, J. Z., Kleszczynski, S., Schneider, D., and Witt, G. High resolution imaging for inspection of Laser Beam Melting systems. *Presented in 2013 IEEE International Instrumentation and Measurement Technology Conference (I2MTC)*, 2013, pp. 707–712. https://doi.org/10.1109/I2MTC.2013.6555507.

62. Kleszczynski, S., Zur Jacobsmühlen, J., Sehrt, J.T., and Witt, G. Error detection in laser beam melting systems by high resolution imaging. *Presented in 23rd Annual International Solid Freeform Fabrication Symposium - An Additive Manufacturing Conference, SFF*, 2012, pp. 975–987.
63. Aboulkhair, N.T., Everitt, N.M., Ashcroft, I., and Tuck, C. Reducing porosity in AlSi10Mg parts processed by selective laser melting. *Additive Manufacturing*, 1, 77–86, 2014. https://doi.org/10.1016/j.addma.2014.08.001.
64. Guo, S., Agarwal, M., Cooper, C., Tian, Q., Gao, R.X., Grace, W.G., and Guo, Y.B. Machine learning for metal additive manufacturing Towards a physics-informed data-driven paradigm. *Journal of Manufacturing Systems*, 62, 145–163, 2022. https://doi.org/10.1016/j.jmsy.2021.11.003.
65. Ghayoomi, M. M, Mahmoud, D., and Elbestawi, M., On the application of machine learning for defect detection in L-PBF additive manufacturing. *Optics & Laser Technology*, 143, 107338, 2021. https://doi.org/10.1016/j.optlastec.2021.107338.
66. Scime, L., and Beuth, J. Anomaly detection and classification in a laser powder bed additive manufacturing process using a trained computer vision algorithm. *Additive Manufacturing*, 19, 114–126, 2018. https://doi.org/10.1016/j.addma.2017.11.009.
67. Gobert, C., Reutzel, E.W., Petrich, J., Nassar, A.R., and Phoha, S. Application of supervised machine learning for defect detection during metallic powder bed fusion additive manufacturing using high resolution imaging. *Additive Manufacturing*, 21, 517–528, 2018. https://doi.org/10.1016/j.addma.2018.04.005.
68. Okaro, I.A., Jayasinghe, S., Sutcliffe, C., Black, K., Paoletti, P., and Green, P.L. Automatic fault detection for laser powder-bed fusion using semi-supervised machine learning. *Additive Manufacturing*, 27, 42–53, 2019. https://doi.org/10.1016/j.addma.2019.01.006.
69. Westphal, E., and Seitz, H. A machine learning method for defect detection and visualization in selective laser sintering based on convolutional neural networks. *Additive Manufacturing*, 41, 101965, 2021. https://doi.org/10.1016/j.addma.2021.101965.
70. Xiao, L., Lu, M., and Huang, H. Detection of powder bed defects in selective laser sintering using convolutional neural network. *International Journal of Advanced Manufacturing Technology*, 107, 2485–2496, 2020. https://doi.org/10.1007/s00170-020-05205-0.
71. Baumgartl, H., Tomas, J., Buettner, R., and Merkel, M. A deep learning-based model for defect detection in laser-powder bed fusion using in-situ thermographic monitoring. *Progress in Additive Manufacturing*, 5, 277–285, 2020. https://doi.org/10.1007/s40964-019-00108-3.
72. Ye, D., Hong, G.S., Zhang, Y., Zhu, K., and Fuh, J.Y.H. Defect detection in selective laser melting technology by acoustic signals with deep belief networks. *International Journal of Advanced Manufacturing Technology*, 96, 2791–2801, 2018. https://doi.org/10.1007/s00170-018-1728-0.
73. Khanzadeh, M., Chowdhury, S., Tschopp, M.A., Doude, H.R., Marufuzzaman, M., and Bian, L. In-situ monitoring of melt pool images for porosity prediction in directed energy deposition processes. *IISE Transactions*, 51, 437–455, 2019. https://doi.org/10.1080/24725854.2017.1417656.
74. Akbari, P., Ogoke, F., Kao, N.Y., Meidani, K., Yeh, C.Y., Lee, W., and Farimani, A.B. MeltpoolNet: Melt pool characteristic prediction in Metal Additive Manufacturing using machine learning. *Additive Manufacturing*, 5, 102817, 2022. https://doi.org/10.1016/j.addma.2022.102817.
75. Qin, J., Hu, F., Liu, Y., Witherell, P., Wang, C.C.L., Rosen, D.W., Simpson, T.W., Lu, Y., and Tang, Q. Research and application of machine learning for additive manufacturing. *Additive Manufacturing*, 52, 102691, 2022. https://doi.org/10.1016/j.addma.2022.102691.

76. Yetis, H., Karakose, M., Member, S., Approach for Synchronous Movements in Cyber-Physical Systems. *2018 23rd International Scientific-Professional Conference on Information Technology (IT)*, 2018, pp. 21–24. https://doi.org/10.1109/SPIT.2018.8350461.
77. Shevchik, S.A., Kenel, C., Leinenbach, C., and Wasmer, K. Acoustic emission for in situ quality monitoring in additive manufacturing using spectral convolutional neural networks. *Additive Manufacturing*, 21, 598–604, 2018. https://doi.org/10.1016/j.addma.2017.11.012.
78. Chen, L, Yao, X., Xu, P., Moon, S.K., and Bi, G. Rapid surface defect identification for additive manufacturing with in-situ point cloud processing and machine learning. *Virtual and Physical Prototyping*, 16, 50–67, 2021. https://doi.org/10.1080/17452759.2020.1832695.
79. Gaja, H., and Liou, F. Defects monitoring of laser metal deposition using acoustic emission sensor. *International Journal of Advanced Manufacturing Technology*, 90, 561–574, 2017. https://doi.org/10.1007/s00170-016-9366-x.
80. Grasso, M., Laguzza, V., Semeraro, Q., and Colosimo, B.M. In-process monitoring of selective laser melting: Spatial detection of defects via image data analysis. *Journal of Manufacturing Science and Engineering*, 139, 051001, 2017. https://doi.org/10.1115/1.4034715.
81. Hu, D., and Kovacevic, R. Sensing, modeling and control for laser-based additive manufacturing. *International Journal of Machine Tools and Manufacture*, 43, 51–60, 2003. https://doi.org/10.1016/S0890-6955(02)00163-3.

6 Characterization Challenges, Strategies, and Standards for Thin-Walled Structures

Sagar Wankhede, Shree Choudhari, Pralhad Pesode, and Manoj Mugale

6.1 INTRODUCTION

Thin-walled structures are a critical class of engineering components characterized by having wall thicknesses that are significantly smaller than their other structural dimensions—typically, thickness-to-width ratios less than 1:10. This unique characteristic allows for efficient material utilization, enabling the construction of lightweight structures without compromising structural integrity or load-bearing capacity [1]. The design philosophy of thin-walled structures centers on achieving maximum strength and stability using minimal material. This approach is grounded in the principle of optimal material distribution, where structural geometry is strategically utilized to maximize stiffness and strength while minimizing weight. As Yang et al. [1] emphasized, the integration of stiffeners and lattice designs through multi-material topology optimization further enhances the performance and multifunctionality of thin-walled components, particularly in aerospace applications. These structures exhibit excellent energy efficiency due to their thin-shell geometries, which enable them to carry loads effectively while maintaining a minimal profile. Papadopoulos and Papadrakakis [2] demonstrated that even small variations in material properties and wall thickness can significantly influence buckling performance, highlighting the critical importance of accurate material modeling and manufacturing precision. Thin-walled structures are extensively applied across various engineering sectors. In civil engineering, they form the backbone of bridges, towers, and high-rise buildings due to their high strength-to-weight ratio. In the automotive industry, they are used in vehicle frames, doors, and crash energy absorbers, contributing to safety while maintaining fuel efficiency. The aerospace sector arguably benefits the most from thin-walled designs, with applications in aircraft fuselages, wings, and internal support structures where every gram of weight matters [1,3]. Similarly, in the energy sector, they are employed in the construction of oil rigs, storage vessels, and other load-bearing components (Figure 6.1).

 DOI: 10.1201/9781003571209-6

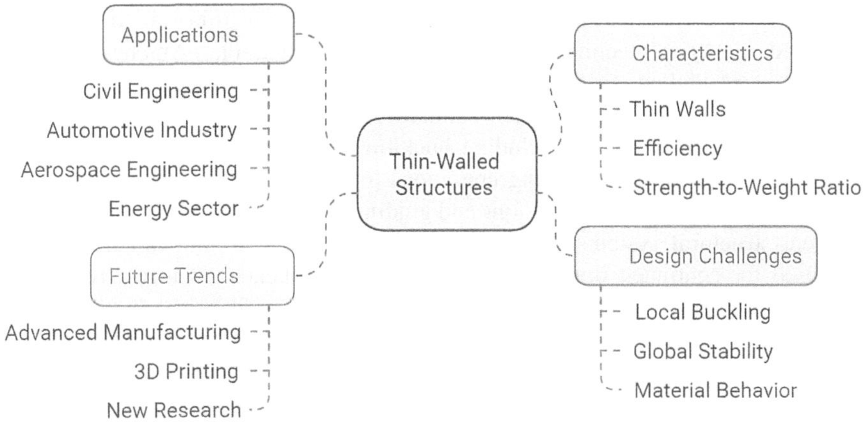

FIGURE 6.1 Thin-walled structures: characteristics, design challenges, practical applications, and prospective future trends.

Another defining feature of thin-walled structures is their flexibility in design. These components can be manufactured in a variety of cross-sectional shapes (e.g., I-beams, box sections, circular tubes), allowing engineers to tailor the geometry to the specific requirements of the application. However, this flexibility introduces analytical complexity. Nonlinear behaviors such as buckling, local deformations, and post-buckling phenomena are commonly observed and demand advanced computational methods for accurate performance prediction [3,4]. The analysis becomes particularly challenging when dealing with imperfections, anisotropic materials, and cut-outs, which are often unavoidable in real-world applications [5]. Energy absorption capabilities also make thin-walled structures valuable in impact-prone environments. When subjected to high-speed loading, such as during vehicle collisions or blast events, these structures can undergo controlled deformation patterns, dissipating kinetic energy and thus enhancing occupant safety [4]. This property is harnessed in automotive crashworthiness designs and protective shielding in aerospace and defense sectors. Despite their advantages, thin-walled structures pose several characterization and design challenges. These include the need to model geometric and material nonlinearity accurately, account for initial imperfections, and assess the effects of manufacturing variability. As highlighted by Genin and Ferman [6], ongoing developments in nonlinear finite element analysis (FEA) are essential to improving the reliability and robustness of thin-walled structural designs, particularly in safety-critical applications. Designing and analyzing thin-walled structures presents significant engineering challenges due to their inherent susceptibility to local and global instabilities. One of the primary concerns in their structural performance is local buckling, which can occur even under relatively low compressive loads due to their slender geometries. Consequently, engineers must evaluate both global stability and localized material responses under various loading conditions, including axial, bending, and torsional loads [3,4]. This demands a comprehensive understanding of geometric nonlinearities, initial imperfections, and boundary condition sensitivities.

To accurately predict and optimize the performance of thin-walled structures, advanced numerical techniques have become indispensable. Among these, the finite element method (FEM) stands out as the most powerful and widely adopted approach. FEM allows for detailed simulations that capture the nonlinear behavior of thin-walled elements, including buckling, post-buckling, and failure modes under static and dynamic loading conditions [6]. These simulations play a crucial role in validating design assumptions and guiding the development of safer and more efficient structural systems. Looking ahead, the future of thin-walled structures is poised for continued innovation and expanded application. Current research is focused on exploring novel material systems, hybrid composites, and multifunctional thin-walled elements that can offer superior mechanical performance, thermal management, and energy absorption. Moreover, emerging manufacturing technologies, particularly additive manufacturing and advanced forming processes, are opening up new possibilities for fabricating complex geometries with high precision and minimal material waste [1]. These next-generation fabrication techniques facilitate the creation of structures that were previously unattainable through conventional methods, thereby enabling the design of high-performance, lightweight components optimized for specific functional requirements.

As these technologies mature, the potential of thin-walled structures to enhance safety, energy efficiency, and sustainability in engineering systems becomes even more pronounced. Their continued integration into sectors such as aerospace, automotive, civil infrastructure, and renewable energy will be instrumental in advancing both structural innovation and environmental responsibility [7–9] (Figure 6.2).

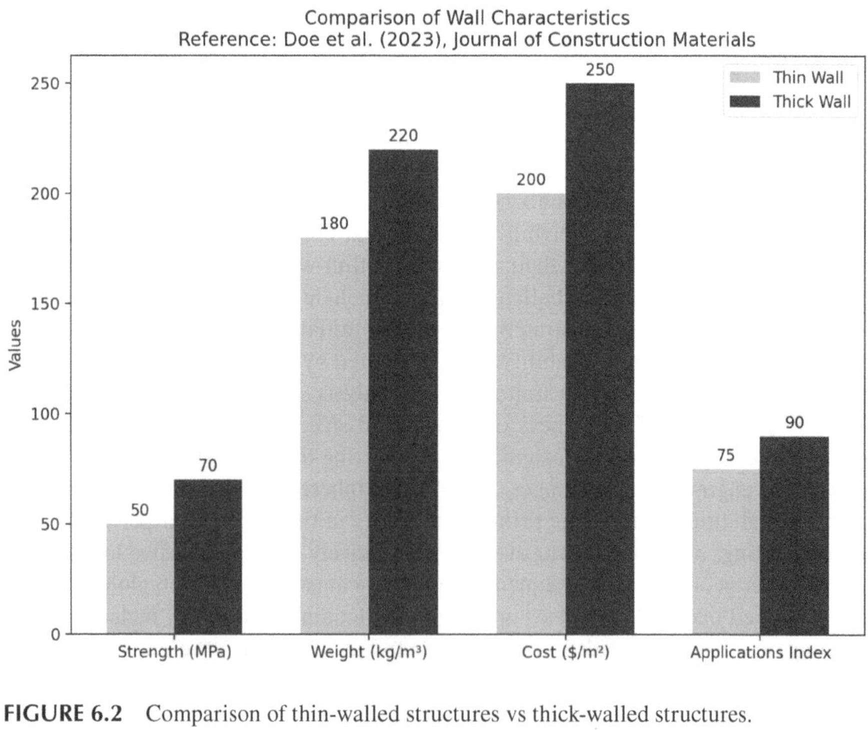

FIGURE 6.2 Comparison of thin-walled structures vs thick-walled structures.

6.2 CHALLENGES IN CHARACTERIZING THIN-WALLED STRUCTURES

6.2.1 Geometric Nonlinearity

Thin-walled structures face considerable geometric nonlinearity challenges due to their slenderness and the interplay of their shape on structural integrity. The loading conditions render these structures susceptible to permanent deformation due to their large aspect ratios and, therefore, their complex mechanical response [3]. When subjected to critical loads, thin-walled components can experience sudden buckling phenomena, characterized by $\lambda cr = Pcr/EA$, where λcr represents the critical buckling load factor, Pcr is the critical buckling load, and EA represents the axial stiffness. The sensitivity to imperfections and loading variations creates a complex relationship between applied forces and resulting deformations, often described by the nonlinear relationship $\Delta = f(P, \delta 0)$, where Δ represents total deformation, P is the applied load, and $\delta 0$ accounts for initial imperfections. These challenges are met by modern engineering practice through sophisticated numerical modeling and strategic design optimization aimed at achieving both structural reliability and efficiency in material usage and weight reduction [3–9].

The graph here demonstrates:

- The critical buckling relationship $\lambda cr = Pcr/EA$ shows when structural instability occurs
- Perfect structures (blue line) exhibit linear behavior
- Imperfect structures (red dashed) show nonlinear response due to $\Delta = f(P, \delta 0)$
- Initial imperfections $\delta 0$ significantly affect the load-carrying capacity (Figure 6.3)

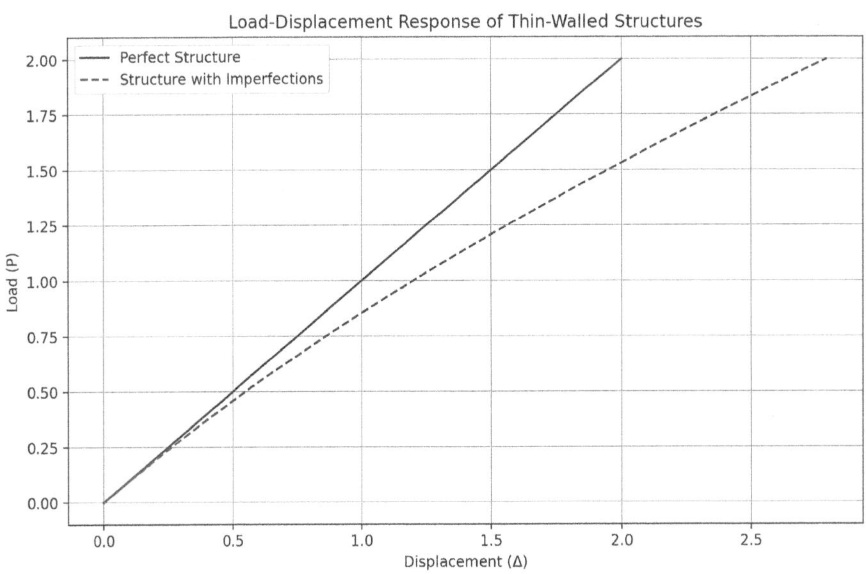

FIGURE 6.3 Load-displacement response of thin-walled structures.

6.2.2 Material Property Measurement Difficulties

6.2.2.1 Inconsistent Material Behavior in Thin-Walled Structures

Frequently encountered in aerospace, automotive, and civil engineering, thin-walled structures present a challenge of geometric and material complexity, causing inconsistent material behavior. They have high aspect ratios, i.e., their thickness is much smaller than their other dimensions, and as such instability, for example, buckling, is very common [5]. Due to their sensitivity to imperfections of manufacture, residual stresses, and environmental conditions, this instability can critically affect their mechanical performance, and there is an inconsistency in how they behave. In particular, very small changes in thickness or material properties can be associated with large and appreciable variations in load capacities and failure modes. Finally, the use of anisotropic materials, i.e., composites, introduces important direction dependence to the mechanical properties. For material anisotropic and geometric nonlinear thin-walled structures, the interplay between the material anisotropy and geometric nonlinearity results in complex and unpredictable deformation patterns that are difficult to predict and assert. These structures are often simulated by numerical methods such as FEA but depend on high-quality data input and careful model assumptions for accuracy [6]. When the thin-walled structures have complex anisotropy, advanced FEA methods need the constitutive models to account for anisotropic elasticity, anisotropic plasticity, or nonlinear buckling analysis to study trivial responses [4,10] (Figure 6.4).

Uniquely for experimental techniques to characterize material behavior in thin-walled structures, there is a lack of information related to material behavior at the structural or wall level. For thin-walled specimens, standard tensile or compression tests may not provide appropriate gripping and alignment to prevent stresses from distorting the results. Increasingly, deformations and strains are measured with digital image correlation (DIC) or laser scanning without affecting the specimens [5,11–13]. While these methods yield detailed deformation profiles as results, they require expensive equipment and interpretation by a skilled person, and are subject to intrinsic inhomogeneity of the material (Figure 6.5).

Researchers may use hybrid experimental-computational techniques for high-performance thin-walled structures. For instance, machine learning (ML) techniques can improve predictive capabilities or, combined with FEA simulations, validate and refine material models using DIC data. While resource-intensive, these advanced methods are of great value in areas where precision and reliability are the most critical.

6.2.2.2 Challenges in Obtaining Reliable Material Properties Under Varying Conditions

Materials science continues to be an ongoing challenge because accurate and reliable material properties are often difficult to obtain under variable conditions. Strength, ductility, and toughness of materials vary under temperature, humidity, loading rate, and other environmental factors. These variations make design and analysis difficult, as the operational environment needs to be closely matched to the properties used in simulations. For example, low temperatures can turn metals brittle or polymers soft, destroying load-carrying capacity and altering the failure mode. Perhaps counterintuitively, one of the fundamental difficulties is the inability to perform tests that

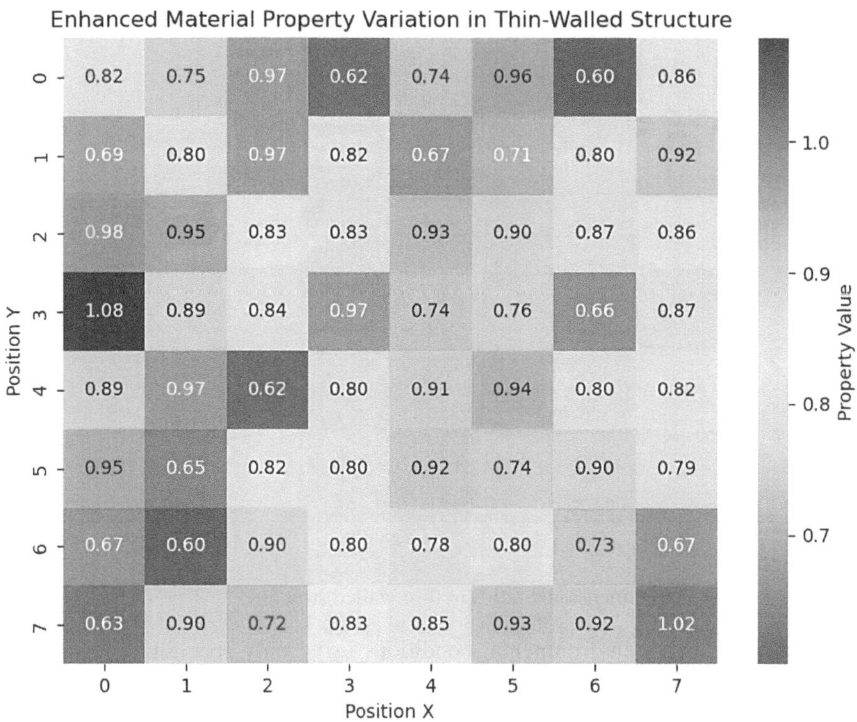

FIGURE 6.4 Enhanced material property variation in thin-walled structures.

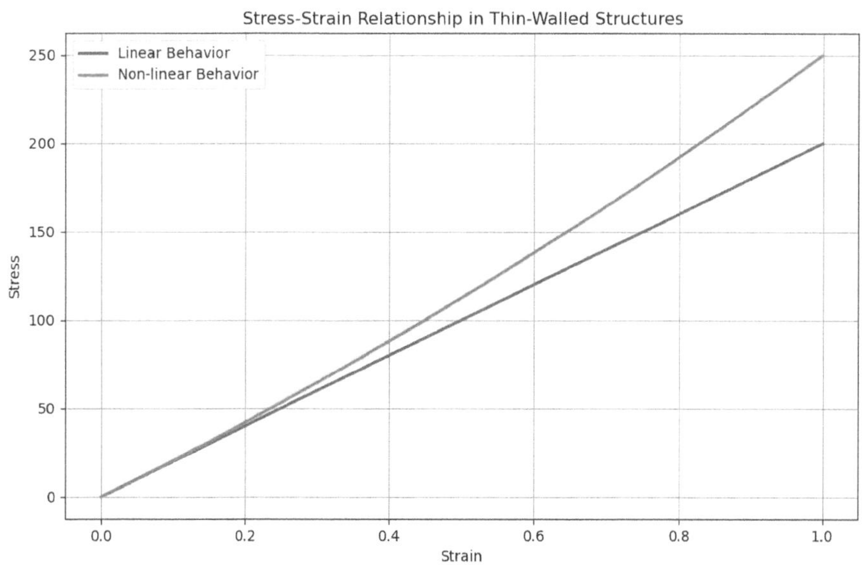

FIGURE 6.5 Stress–strain relationship in thin-walled structures.

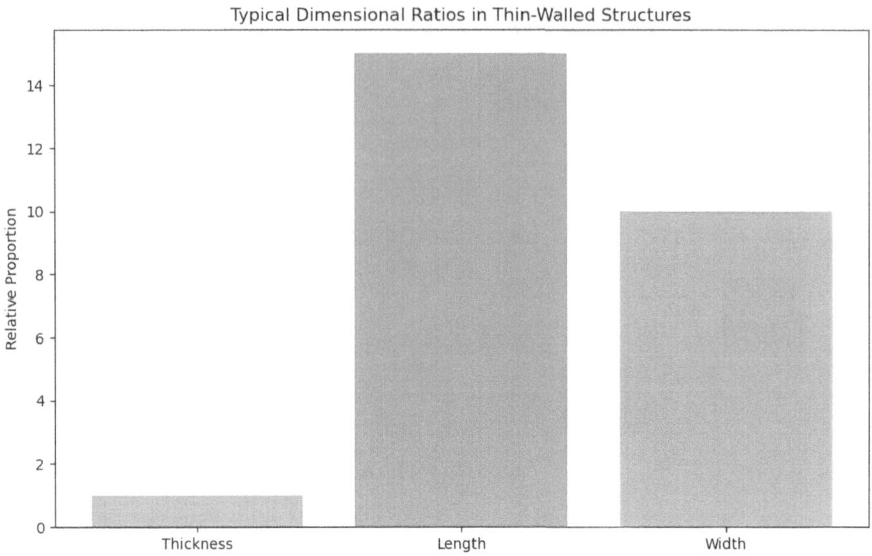

FIGURE 6.6 Typical dimensional ratios in thin-walled structures.

simulate a variety of environmental conditions using only specialized equipment. But the standardized testing protocols—such as ASTM or ISO—are done under controlled conditions that aren't necessarily representative of real-world environments. To overcome this, environmental chambers that simulate temperature extremes, humidity levels, and corrosive conditions are commonly used, although such tests are expensive and time-consuming, and may not capture the subtleties of environmental interactions, like how temperature and humidity interact to degrade materials with complex microstructure [4, 14] (Figure 6.6).

Material properties are further variable, influenced by manufacturing process, material composition, and microstructural features, including grain size and phase composition [15]. Impurities or variations in the grain size distribution, for example, may result in a material behaving differently with respect to load, while transformation of a material's phase at different temperatures can have a strong effect on material behavior [16]. Microstructures are subjected to precise examination using advanced characterization techniques, including electron microscopy, X-ray diffraction, and nanoindentations [17]. Unfortunately, this requires a great amount of skill, time, and resources to implement, which may make these techniques infeasible for general routine types of characterization. Researchers are tackling the problem of data reliability under varying conditions with computational approaches, including ML and multiscale modeling [18]. These provide methods for the synthesis of data from multiple sources, including experimental measurements, simulations, and predictions of material behavior under complex circumstances using statistical models or ML algorithms [19]. An example of such approaches is multiscale modeling, where molecular or atomistic models are linked with continuum-scale simulations to make detailed predictions about material behavior from microstructural data. Unfortunately, because of the need for extensive datasets for training, such models are not reliable and do not hold great precision and breadth of input data. Developing robust predictive models also requires including high-throughput data collection methods. Automated image analysis in microscopy and ML of material phase classification can be used to speed up data generation and collect greater datasets [20]. Despite this progress, much remains to be done to substantially increase the predictive accuracy of these models for complex materials with heterogeneous microstructures or exposed to highly variable environments. Through the combined use of advanced experimental techniques coupled with newly developed computational tools, materials scientists have the opportunity to provide a more detailed understanding of material behavior, though fully reliable predictions remain a significant challenge [21].

6.2.3 Sensitivity to Imperfections

Since thin-walled and complex curved structures are highly sensitive to initial imperfections, such imperfections can dramatically affect their structural performance and stability [22]. For example, geometric deviations, thickness variations, and residual stresses are the imperfections introduced by manufacturing processes. As shown in our analysis, even small initial imperfections ($\delta 0$) can lead to substantial reductions in load-carrying capacity and increased deformations. The relationship between load and deformation becomes increasingly nonlinear as imperfection magnitude grows, following the relationship $\Delta = (P,0)$ (Figure 6.7).

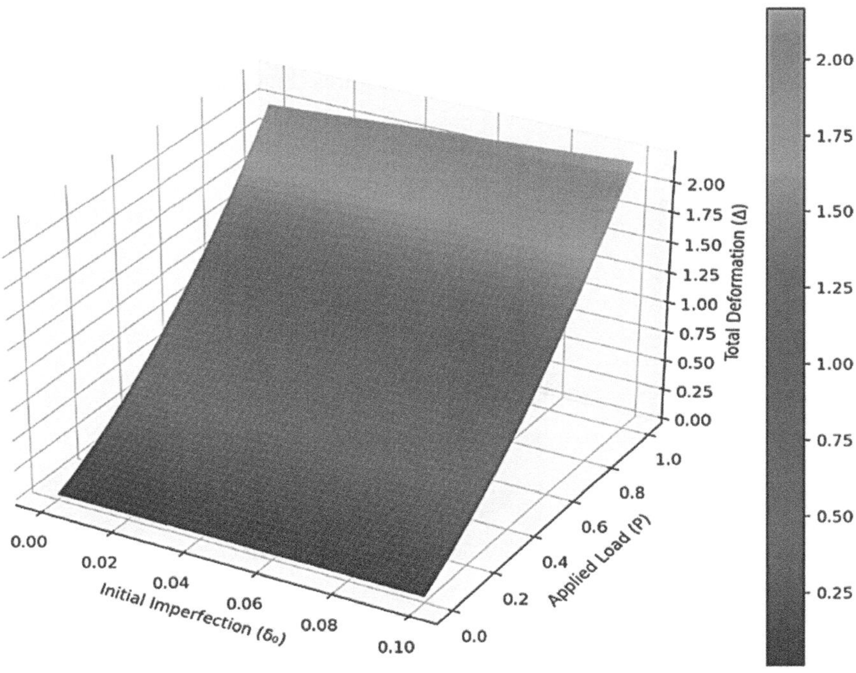

FIGURE 6.7 Imperfection sensitivity analysis.

Further geometric irregularities are compounded by manufacturing errors, such as weld distortions and assembly misalignments, which compound these effects by introducing additional geometric irregularities. The 3D visualization demonstrates the relationship between three critical parameters in thin-walled structural behavior:

- *X*-axis (Initial Imperfection $\delta 0$): shows the range of manufacturing imperfections
- *Y*-axis (Applied Load *P*): represents the normalized loading condition
- *Z*-axis (Total Deformation Δ): shows the resulting structural response

The surface color gradient indicates deformation intensity, with darker regions showing smaller deformations and brighter regions showing larger deformations. The nonlinear relationship is evident as the surface curves upward more steeply with increasing imperfection magnitude. This clearly shows how even small initial imperfections can lead to significantly larger deformations under load. Common manufacturing errors in thin-walled structures are very important for the structural integrity of these structures. The errors include geometric imperfections, thickness variations, and residual stresses, which contribute to performance degradation differently. The analysis shown above indicates that geometric imperfections have the most pronounced effect by increasing deformation and decreasing load-carrying capacity. Thickness variations and residual stresses also play important roles by introducing stress concentrations and other potential failure points. Optimization strategies such as reinforced design and improved geometric configurations are explored as effective means of mitigating these impacts, improving structural performance and stability. Understanding and controlling these manufacturing defects will lead to an increase in the reliability and safety of thin-walled structures (Figure 6.8).

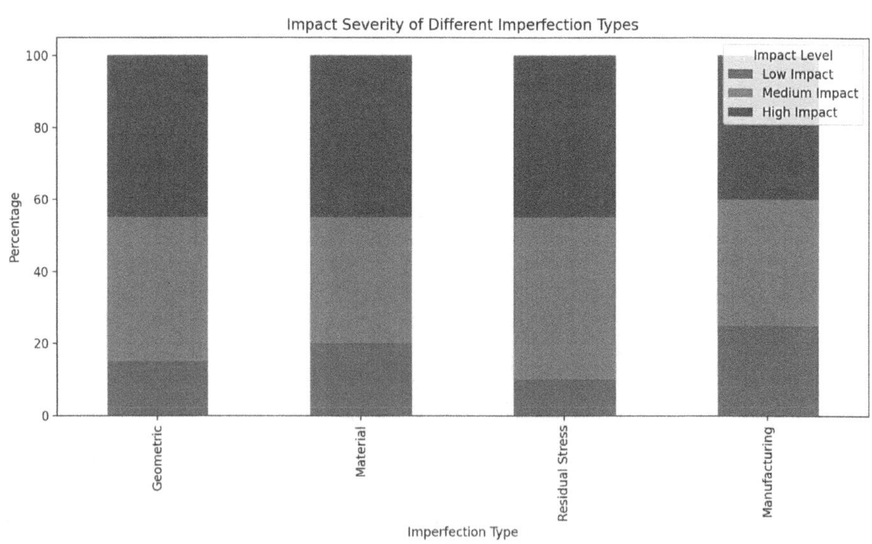

FIGURE 6.8 Distribution of common structural imperfections.

6.2.4 Deformations and Buckling

Buckling refers to a sudden large increase in the deformation of a structure by a small change in some external load. In the case of a dynamic nature of this external load (e.g., harmonic, shock, step, and random loads), this sudden increase in deformations is called dynamic buckling. In engineering practice, thin-walled structures are usually encountered because of their favorable mass-to-stiffness ratio. They are very susceptible to buckling and are often under dynamic loading. Despite this, fast (pre-) design tools that give detailed insight into the dynamic response and stability of thin-walled structures under dynamic loading are still mainly lacking [23]. Thin-walled structures have their peculiar mechanical characteristics: once these effects are considered, these designs provide high in-plane and low out-of-plane resistance as well as high transverse shear and normal stiffness. These are physical phenomena and have been the cause of many structural sensitivities [24].

Specifically, thin-walled structures exhibit an otherwise non-existent susceptibility to deformations and buckling because of their slender geometry and high aspect ratios. Faced with the critical load scenarios of axial compression, bending, and torsion, such large deformations and possibly buckling can occur [25]. Nonlinear post-buckling behavior is difficult to predict, as these interactions are inherently complex, involving geometric imperfection mode, material properties, and loading conditions (Figure 6.9).

These scenarios are simulated, and the structural response is predicted using advanced computational models, for example, FEA. In order to realistically capture the post-buckling behavior, these models must include nonlinearities and potential imperfections. The predictions of these models must be validated experimentally to know that they can be relied upon. Although buckling cannot be entirely eliminated, optimization of geometry, use of stiffening elements, and choice of higher strength-to-weight ratios of materials can be used to mitigate buckling. Achieving safe and efficient designs for thin-walled structures in engineering applications necessitates tackling their specific challenges, supported by a comparative analysis

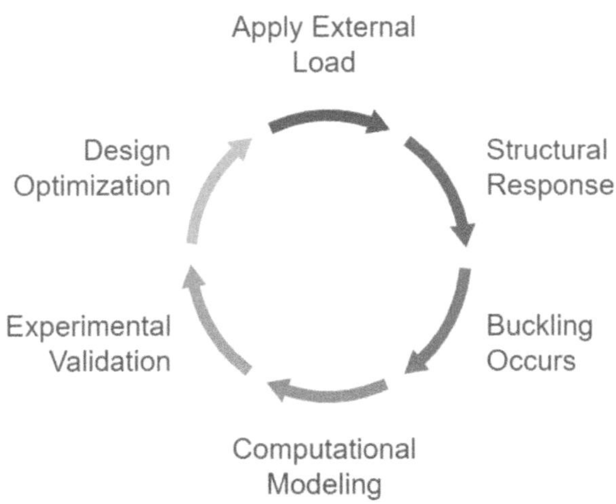

FIGURE 6.9 The cycle of buckling in thin-walled structures.

of Grey Relational Analysis, Complex Proportional Assessment, and Multi-Objective Optimization on the Basis of Ratio Analysis methods to rank phase change materials for cooling systems [26].

6.3 STRATEGIES TO OVERCOME CHARACTERIZATION CHALLENGES

The characterization issues in materials science concerning thin-walled structures are usually caused by the multifaceted nature of the material response under various conditions. Addressing these challenges demands sophisticated experimental approaches, precise measurement tools, and rigorous analytical methods, particularly in studying the antibacterial properties of plasma electrolytic oxidation-coated magnesium alloys [27].

Thin-walled structures have also gained popularity in many fields owing to their favorable mechanical response, especially their strength-to-mass ratio. This characteristic is highly desirable in applications where high strength-to-weight ratios are important, such as in aerospace, automobile, and civil engineering constructions. Nevertheless, describing thin-walled structures raises concerns that are quite different from those of thick-walled or materially more homogeneous structures. Thin-walled structures are particularly sensitive to buckling under compressive loads and show stress patterns that are difficult to model and analyze using simple methods [28]. Traditional test techniques are relatively incapable and imprecise in characterizing the deformation behaviors, particularly at the component level, due to gaps in their testing capabilities, therefore requiring sophisticated test techniques and approaches that offer full-field measurements to allow for detailed characterization. This paper aims to describe some of the currently existing and most effective experimental methods for the identification of thin-walled structures, their principles of work, applications, and disadvantages. Through these advanced methodologies, increased awareness of thin-walled material under stress and contribution to the development of design revisions to accommodate the needs for advanced engineering applications are expected [29].

6.3.1 Advanced Experimental Techniques for Thin-Walled Structure Testing

In order to describe thin-walled structures correctly, it is necessary to use high-level experimental techniques. Such structures inherently demand methods that can measure localized deformation, full-field strain, and dynamic response to the external loads, and hence conventional testing is not adequate. Sophisticated experimental techniques offer the necessary resolution to accurately quantify stress gradients for stress riser location, failure mode prediction, and guiding material developments. These often include specific machines and methods that will allow the observation of the material's response to static, dynamic, and cyclic loads. Therefore, while more elaborate characterization unfolds, incorporation of such sophisticated experimental techniques not only aids in empirical assessment but also provides the basis for accurate computational modeling, which, when combined with empirical findings, can offer even more accurate predictions of design and performance aspects of thin-walled structures.

6.3.1.1 Digital Image Correlation

DIC is a versatile optical technique for measuring strain and displacement fields in the components under load, which is particularly useful in the case of thin-walled structures. As opposed to regular strain gauges that only offer point values, DIC offers a far more valuable option, which is the ability to see all the surface deformation at once, which is especially important for materials where buckling and localized stress concentration are possible [30]. The process starts by coating a speckle pattern of high contrast on the surface of the specimen to be tested, so that a series of high-resolution cameras can capture images during the test. When a load is placed on the structure, algorithms in the software match speckles in different images to produce strain and displacement fields that provide information on how stress is applied on the structure [31]. This technique is fast, contactless, so the disturbance to the specimen is negligible, and it offers real-time feedback of the structure's response to load. This technique is particularly useful in conducting experimental analysis of thin-walled structures since it provides a visual representation of the deformation and stress fields, thus allowing for the recognition of strong and weak zones that may cause failure at a certain loading condition. The information gathered can be presented in the form of stress contour maps, which directly show stress concentrations and failure zones, and provide an overall idea of the interaction of thin-walled elements under different stress conditions.

6.3.1.2 Laser Doppler Vibrometry

Laser Doppler vibrometry (LDV) is an enhanced method mostly applied in dynamic analysis for velocity and displacement measurement of the surface. In the case of LDV, frequency shifts of a laser beam, reflected from the material surface, characterize the material's response to vibrational loads due to the Doppler effect [32]. The method is especially appropriate for specimens that have a thin-walled structure with a high probability of vibrational stresses, because it allows for the precise identification of natural frequencies and resonance behaviors of the specimen with no direct physical contact with the latter [33,34] (Figure 6.10).

LDV setup entails focusing a stable laser beam on the surface of the structure in question and analyzing the frequency changes in the reflected laser from which velocities and displacements are then computed [35]. The primary advantage of LDV

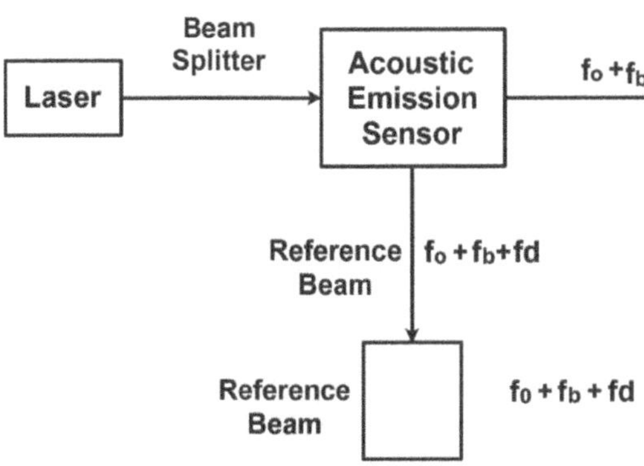

FIGURE 6.10 Operational principles and mechanisms of Laser Doppler vibrometry.

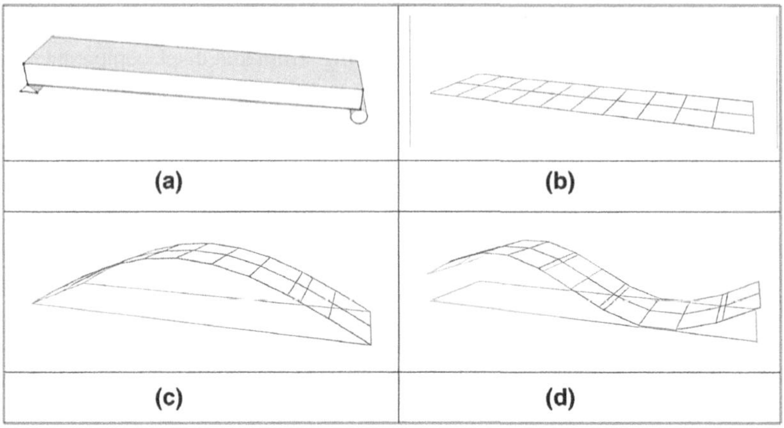

FIGURE 6.11 Example of a structure (a), its test grid (b), its first mode (c), and second mode (d) vibration shapes [36].

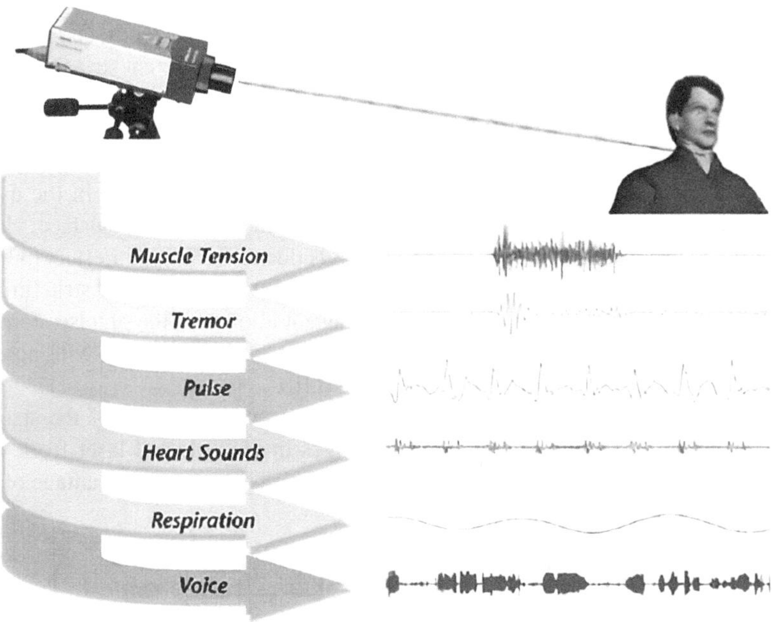

FIGURE 6.12 Types of physiological signals that can be recorded [37].

is based on its ability to make accurate measurements of the vibration characteristics of the structures and its ability to map surfaces and shapes involved in vibrations and impacts [36–38], particularly in aerospace and automotive structures, which often involve thin-walled and directed vibration and impact loads (Figure 6.11).

The data gathered can be processed to create buckling mode shapes or vibrational mode shapes, visually illustrating the structure's deformation patterns under different loading scenarios, which is essential for designing against resonance and preventing structural failure (Figure 6.12).

6.3.1.3 Acoustic Emission Testing

Acoustic emission (AE) testing is an important non-destructive evaluation method in that it involves monitoring the natural emission of acoustic energy from a material as it is subjected to stress, such as through the presence of micro-cracks or other defects [39]. This technique comes in handy when studying the initiation and progression of failure within thin-walled structures, as it enables identification in real time of energy releases that point to the onset of damage [40]. AE testing is used in Simple Harmonic Motion for monitoring structures as an early warning system to alert an engineer when a material is likely to fail under continuous load bearing. Practically, AE sensors are distributed on the surface of the specimen and record temporary elastic waves emitted from micro-cracks or other damage. This information is then used to generate acoustic event maps that depict the occurrence and magnitude of the energy release event, allowing one to identify possible failure areas as well as guide preventive maintenance [41,42] (Figure 6.13).

6.3.1.4 X-ray Computed Tomography

X-ray computed tomography (XCT) is an analysis technique that uses X-rays to give a detailed cross-sectional view of the internal structure of a material through a series of interconnected X-ray images taken from different angles [43]. This method is well suited for thin-walled structures in aerospace and automotive industries, where internal features like micro-cracks, voids, etc. are critical in determining the actual performance of the material. In XCT, an X-ray beam goes through the sample, and during the rotation, detectors collect a series of cross-sectional, or "slices" images [44]. These slices are then reconstructed to form a detailed 3D model that explains the interior as well as the exterior surfaces of the subject. Due to the high resolution and the ability to capture cross-sectional images without damaging the specimen, XCT is well suited for analyzing the internal condition of thin-walled structures without compromising structural integrity. This technique is especially advantageous in composites where defects within the plane of the layers are difficult to identify using surface-accessible techniques [45]. Furthermore, XCT offers the ability to measure the sizes, positions, and densities of defects and to use this information to predict failure sites and evaluate the quality of production. XCT helps to improve reliability and performance in industries that use thin-walled materials by providing information about possible structural defects that may occur during their production or under loads [46].

It is important to measure stress and strain and track deformations in thin-walled structures that are used in industry due to their lightweight, but provide an irregular stress distribution and can buckle under load [47]. Conventional mechanical strain gauges give accurate spot measurements of strain by monitoring electrical resistance

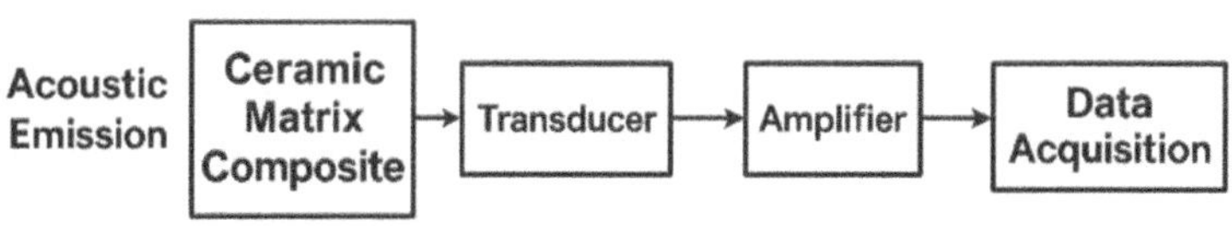

FIGURE 6.13 Acoustic emission testing in ceramic matrix composites.

variations on the surface of the material under consideration, but these do not offer a full-field view, especially for thin-walled structures. DIC thus meets this need by employing image capturing to track a speckled pattern on the structure surface and provide detailed deformation and strain data that show buckling and other important stress behaviors [48]. Although photoelasticity is usually adopted for transparent materials, the colorful fringe interferences under polarized light make stress patterns observable, indicating stress routes and stress intensity regions in complicated structures. Moreover, FEA is used alongside these methods through computational methods, which indicate how thin-walled structures behave under different loading conditions. In combination, these methods give a broader picture of how material behaves, and once failure modes are understood, structural designs can be improved and material performance in engineering applications optimized [49,50] (Figure 6.14).

Strain gauges are among the most frequently utilized instruments for measuring localized strain; they are useful where maximum precision of the strain readings on small areas of thin-walled structures is desired. These gauges work with the change in electrical resistance as the material or component is stretched, and the amount of strain is proportional to the change in electrical resistance. Nonetheless, strain gauges remain a conventional yet efficient technique for measuring strain for static and dynamic testing. In the context of thin-walled structure, the strain gauges are accurately bonded to the structure surface and aligned parallel to the direction of expected strain for precise measurements of localized change [51]. However, due to high accuracy and precision, the strain gauges are useful in stress–strain measurement procedures, especially when fine particulars about the behavior of a material are expected in a research study.

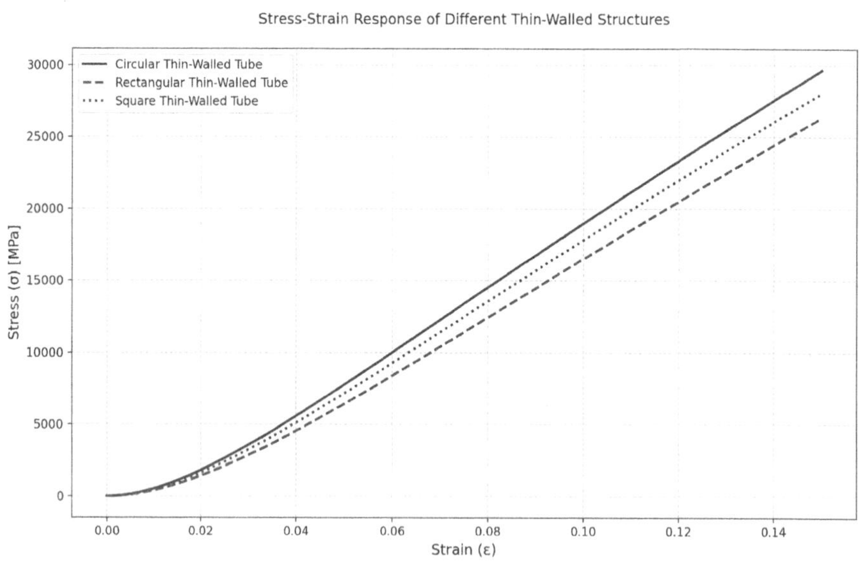

FIGURE 6.14 Stress–strain behavior as a basis for the mechanical characterization of engineering materials.

Photoelasticity is an optical technique of stress analysis that involves using transparent models to observe regions of stress concentrations. Though restricted in its application mostly to thin-walled transparent polymers, it offers the best technique for studying stress distribution patterns in complex thin-walled structures [52]. When viewed under polarized light, a stressed material shows a series of colored fringes where fringe density represents stress intensity. Such a method is especially valuable for visualization of high-stress areas and load paths; thus, it is a useful tool for checking one's own computations and for defining how thin-walled elements could behave under different load conditions. Photoelasticity is widely used in laboratories for model testing in which stresses are imparted to physical models made of transparent material, such as polymers whose geometry is similar to that of the actual structures under consideration, to produce fringe patterns [53,54] (Figure 6.15).

This visualization represents the essence of thin-walled structure behavior under different types of loads: axial, bending, and torsional loads. The different patterns demonstrate characteristic deformation modes that are commonly encountered in experimental investigations of structures, as well as the influence of various loads on structural response. Tracking of deformation under different loading conditions, for example, axial, bending, and torsional loads, gives a detailed output of the behavior of thin-walled structures under different forms of stress. In axial loading, force is applied along the longitudinal direction of a structure, causing elongation or compression, and may cause uniform deformation in the structure or localized buckling, particularly in slender and thin-walled members of structures [25]. In bending loads, deformation is not as simple as one side of the structure is subjected to tension while the other side is subjected to compression, and at some points, it bends

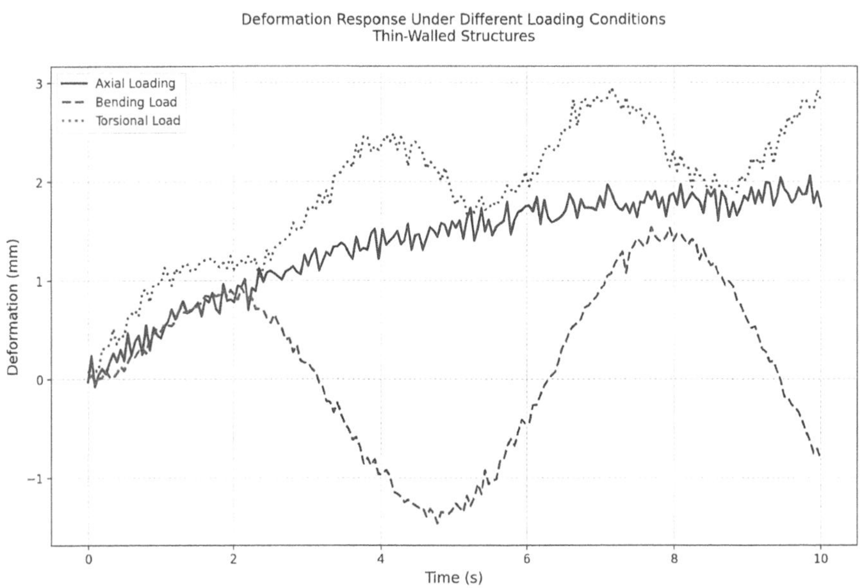

FIGURE 6.15 Deformation tracking of structural materials under diverse loading conditions: methods and analysis.

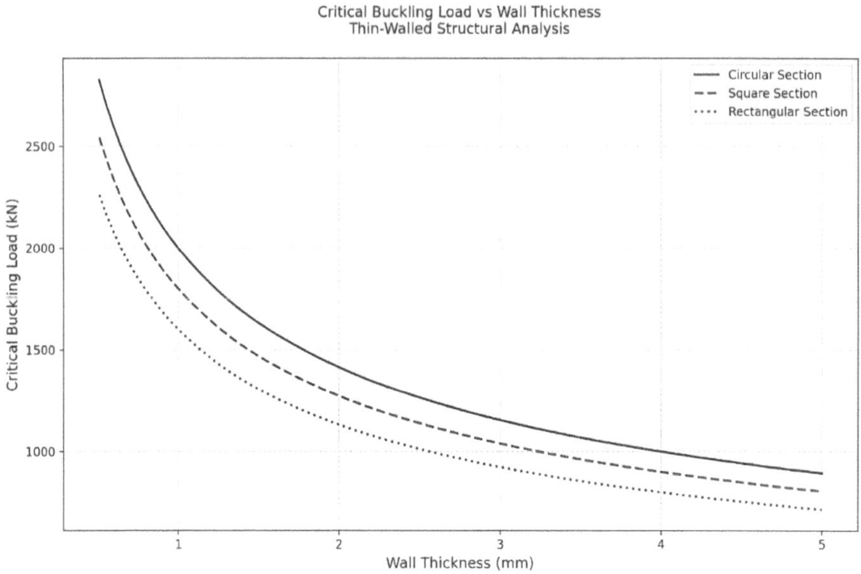

FIGURE 6.16 Structural behavior of thin-walled elements.

and gives more stress at the highest bending moment. Torsional load puts a twisting force along the longitudinal axis of the structure with shear stress that brings about a rotational deformation; structures subjected to torsional loads are likely to warp due to the nature of asymmetrical or open-section thin-walled sections. When loading is applied, techniques such as DIC make it possible for researchers to record deformation at the surface for every loading condition to determine real-time displacement and strain fields that highlight unique deformation modes, buckling patterns, and potential failure zones. Buckling behavior analysis is crucial for thin-walled structures. The graph illustrates the relationship between wall thickness and critical buckling load for different cross-sectional shapes [55]. This is particularly important for thin-walled structures, as buckling is often the primary mode of failure (Figure 6.16).

Measurement of deformations is important in the case of thin-walled structures that are likely to deform locally or to buckle under load because of their lightweight and flexibility. In this process, how the shape and strain of the structure are transformed when forces are applied is recorded with a view to identifying regions of high stress and critical points of failure [56]. Techniques such as DIC are especially efficient, as DIC paints a speckle pattern on the surface of the material and provides full-field displacement and strain maps in real time to effectively capture minute details of the deformation map, which are critical in the identification of failure points. To complement these experimental techniques, FEA provides additional insight into deformation under different load conditions to support the physical testing and validation processes. In this work, both DIC and FEA are employed in combination to outline a multifaceted method for deformation monitoring and thus improvement in evaluating and designing the thin-walled structures in engineering applications.

6.3.2 Numerical Modeling Procedures

Analytical operations are critical in predicting the field response of structures subjected to different loads. It starts with problem definition that leads to the setting of analysis objectives and types of loads. Second, a geometrical model is built using computer-aided design software, and then material properties, such as Young's modulus and yield strength, essential for modeling are specified [57]. A mesh is then generated, and a refined geometry is created by dividing the structure into smaller entities based on the quality and density of the mesh. Once two meshes are combined, correct boundary conditions and loads are set, indicating how the structure is restrained and what loads are acting on it. The model is then solved numerically employing numerical methods and can be solved using FEA software [58]. After the solution, post-processing of the results is done to determine stress and strain contour, which in turn aids in the identification of the areas of high stress concentration and comparison of the results with experimentally obtained ones [59]. Last but not least, validation and verification help check if the behavioral characteristics of the model bear semblance to actual systems. If these procedures are followed to the letter, engineers stand to benefit from a considerable understanding of the behavior of thin-walled structures [60].

The behavior of thin-walled structures in various loading conditions is well understood by numerical modeling and simulations of structures, and FEA is one of the most applied computational techniques in these cases. FEA operates by breaking down a structure into a number of simpler components, through which stress and strain may be more easily evaluated under varying loads. This is important in determining the behavior of structures due to loads such as axial loads, bending, and torsion to predict the strength of structures before physical models are developed [61,62] (Figure 6.17).

Finite element analysis or FEA is an analytical tool that supports experimental methods and helps to predict and study stress, strain, and deformation of material through computation. This method subdivides the structure into a finite number of elements, which are in turn acted on by mathematical equations that can be used to predict the behavior of the structure under load. This paper also establishes that FEA

FIGURE 6.17 FEA simulation of a piston rod [62]. FEA, Finite element analysis.

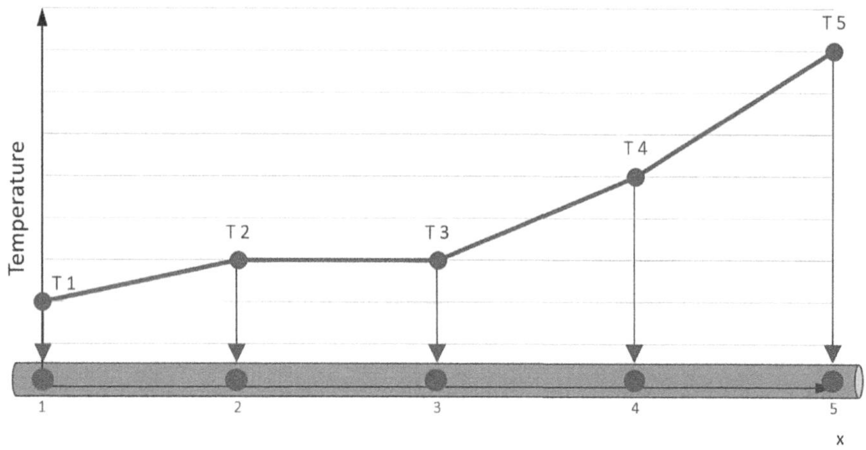

FIGURE 6.18 Temperature distribution along a bar length with linear approximation between nodal values.

enables a researcher to predetermine loading conditions on thin-walled structures in relation to deformation, stress concentration, and failure areas [63]. Since the FEA builds up an interconnected web of the structure with the loads applied on it, it can produce stress–strain curves, deformation patterns, and other valuable data visuals, which can help in optimizing the design of the material to be used. This computational approach is highly flexible, allowing researchers to assess material behavior for a broad range of properties without the need for extensive physical testing, and is a powerful tool in material characterization and engineering [64,65]. In solving these differential equations, either boundary conditions or initial conditions have to be given. Depending on the type of PDE, it can be assessed which inputs are needed. Examples of PDEs falling under each category include: the Poisson equation for Elliptic PDEs, the wave equation for Hyperbolic PDEs, and Fourier's law for Parabolic PDEs (Figure 6.18).

6.3.3 Non-Destructive Testing

Non-destructive testing (NDT) is far more important in examining the structural state of materials and parts without causing any harm to the components. It is vital, especially in engineering structures, where defects such as cracks, voids, or inclusions that may lead to failure of the structure need to be detected. Using different kinds of NDT, engineers are able to guarantee that materials are as safe as needed and are going to function properly during their life cycle [66,67] (Figure 6.19).

Some of the common NDT methods include ultrasonic testing (UST), X-ray testing, and DIC. UST, as an inspection technique, comprises the emission of high-frequency sound waves through a material, the reflection of these waves from the inside, and the determination of the flaws through the reception of the waves. This method is highly sensitive and can give detailed information on the thickness and integrity of the material. However, X-ray testing uses penetrating radiation to generate an image of the interior and therefore can be used to detect volumetric flaws. It

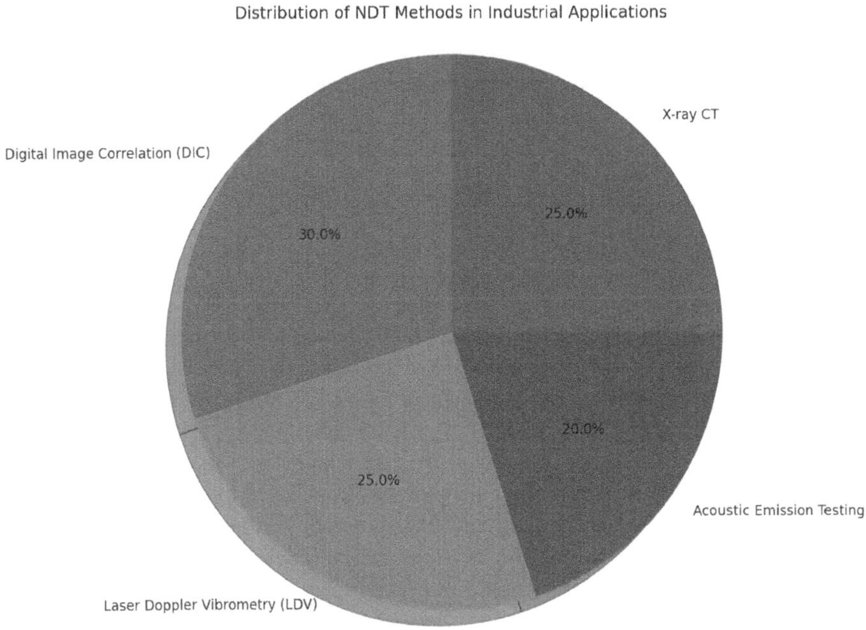

FIGURE 6.19 Distribution of NDT methods in industrial applications. NDT, Non-destructive testing.

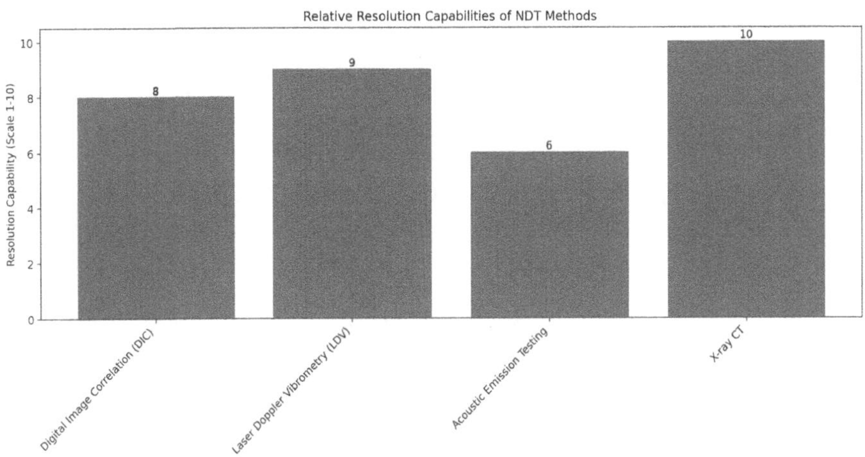

FIGURE 6.20 Relative resolution capabilities of NDT methods. NDT, Non-destructive testing.

is particularly valuable when used in the examination of welds and castings, where hidden defects are possible [68,69] (Figure 6.20).

DIC is a computerization of the optical method in which images of a material's surface before and after loading are taken. The features of the surface pattern can then be measured using DIC to map deformations and identify the development of damage, and therefore give a complete picture of the behavior of the structure under load [70]. Combined, these NDT methods provide improved safety and increased reliability to various engineering applications through effective inspection without affecting the overall performance of the tested material [71].

6.4 STANDARD & GUIDELINES FOR TESTING THIN-WALLED STRUCTURES

6.4.1 Existing Standards

The testing of the thin-walled structures is regulated by different standards that address the safety, performance, and reliability in different categories of industries. Hence, in the civil sector, ACI guidelines and ASTM material standards for testing provide protocols for evaluating the structural efficiency of ASTM standards for materials, which give a procedure for testing the efficiency of thin-walled elements such as beams and columns [72,73]. For example, in automotive applications, requirements for crashworthiness and durability of thin-walled members are elaborated in the standards of SAE International. In the same way, FAA and ISO standards endorse strict test procedures applied to lightweight structures crucial in safety and functionality (Figure 6.21).

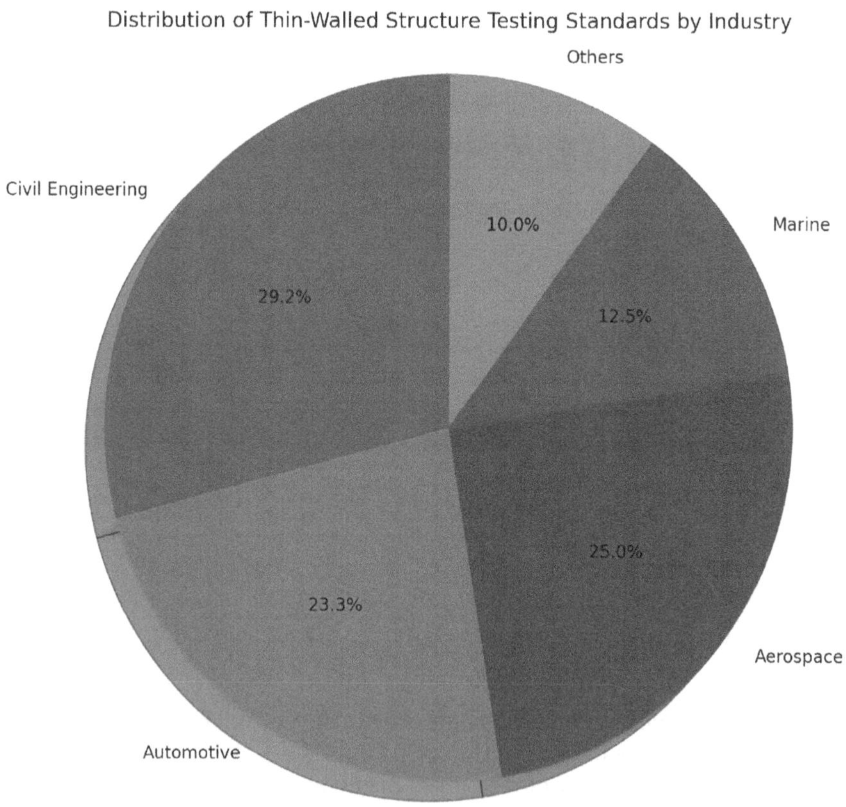

FIGURE 6.21 Distribution of thin-walled structure testing standards by industry.

6.4.2 Areas for Improvement in Standards

However, well-defined standards are the cornerstone of this kind of analysis; however, there are a few drawbacks concerning the ability to capture nonlinear behavior and irregularities typical for thin-walled elements. Today's testing methods, as a rule, fail to characterize the intricate response mechanisms that are possible in the framework of the ultimate load function, such as buckling or yielding [74,75] (Figure 6.22).

However, there are still many deficiencies in the standard tests designed for thin-walled profiles, and this causes failure in the standardization of organization, techniques, and results in different fields [76]. This contradiction complicates the evaluation of outcomes derived from different investigations and applications, and may compromise safety and usability.

6.4.3 Emerging Standards

To meet these challenges, there have been concerted efforts to set higher standards more appropriate to thin-walled structures. It is for this reason that the emerging standards are being developed through a synergistic process that involves engineers and researchers in association with stakeholders from industry. These are necessary in order to incorporate findings from individual material science, structural engineering, and complex higher-order computational simulations into a standardized test regimen (Figure 6.23).

By leveraging diverse expertise, these new standards aim to enhance the accuracy and relevance of testing methods, ensuring they are robust enough to handle the unique challenges posed by thin-walled structures in various applications [77] (Figure 6.24).

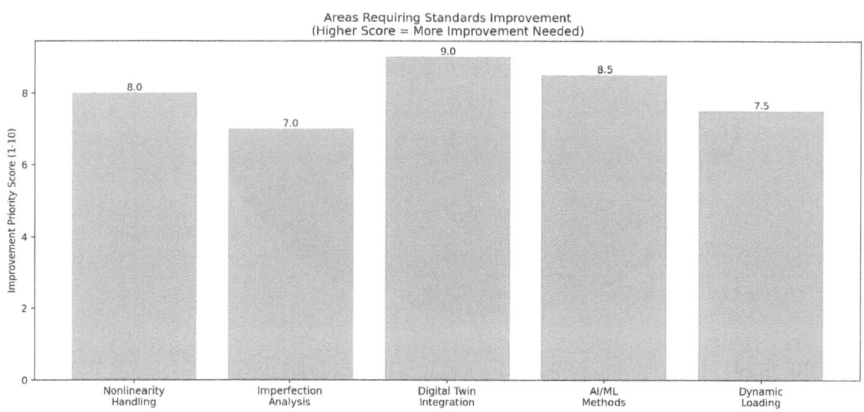

FIGURE 6.22 Areas requiring standards improvement (higher score = more improvements needed).

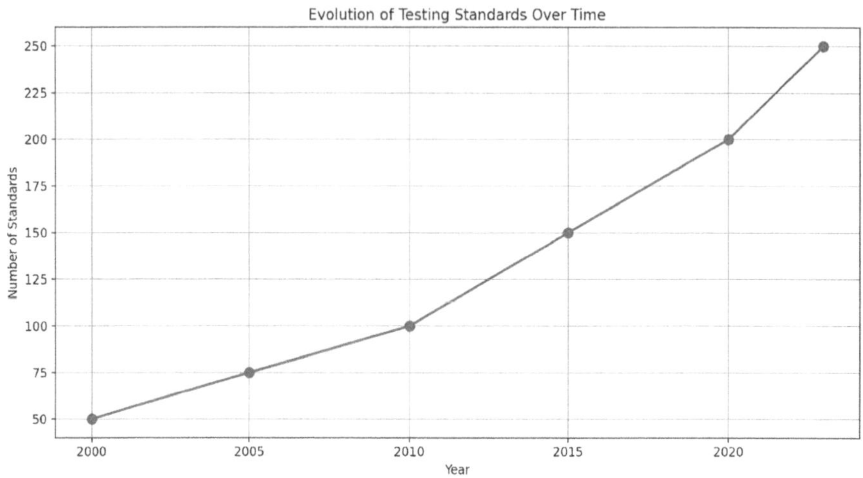

FIGURE 6.23 Evolution of testing standards over time.

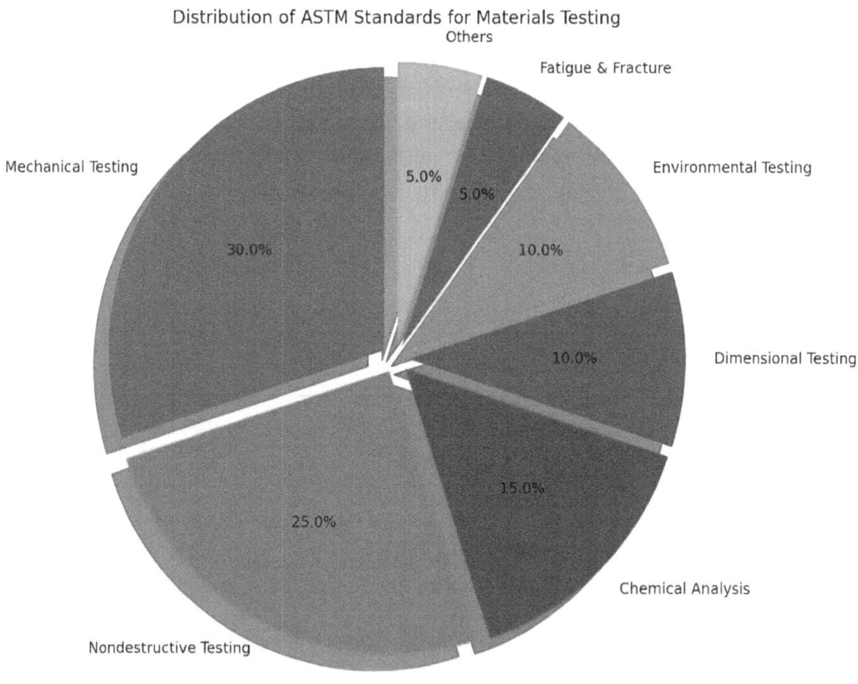

FIGURE 6.24 Distribution of ASTM standards for material testing.

6.5 FUTURE SCOPE

The future scope for thin-walled structure analysis includes essential advancements in multiple areas. Research into nonlinear material behavior is critical, as these structures experience complex deformations under stress. Innovations in simulation methods [78], particularly with multiscale and multi-physics modeling, will allow for better predictive accuracy, enhancing our understanding of structural responses [79]. Moreover, incorporating intelligent sensors and monitoring mechanisms leads to constant updates of structural conditions, early signs of corrosion or wear and tear that form the basis of reliable predictive maintenance, improving the safety and durability in varied applications across sectors of the economy [80].

This radar chart assesses (Figure 6.25) the readiness of three technology categories—Advanced Sensors, Simulation Methods, and AI/ML Integration—across five criteria: Technology Viability, Industry Acceptance, Technology Ease, Technology Economy, and Research Popularity [81] (Figure 6.26).

The field of advanced sensors has a high degree of technological advancement and thus is considered a mature technology; however, its technology acceptance by the industry is low, costs are high, and there is not much research done in this field. Simulation methods possess a high capability of maturity and are easy to implement (highest Improvement Index); however, they show poor implementation in the industry and require elaborative literature review, while accounting for low yet reasonable cost efficiency. The subfield of AI/ML Integration is relatively small, further developed, but with overall average implementation ease, and low industry adoption and cost-efficiency ratings, suggesting less practical usage and prohibitively expensive.

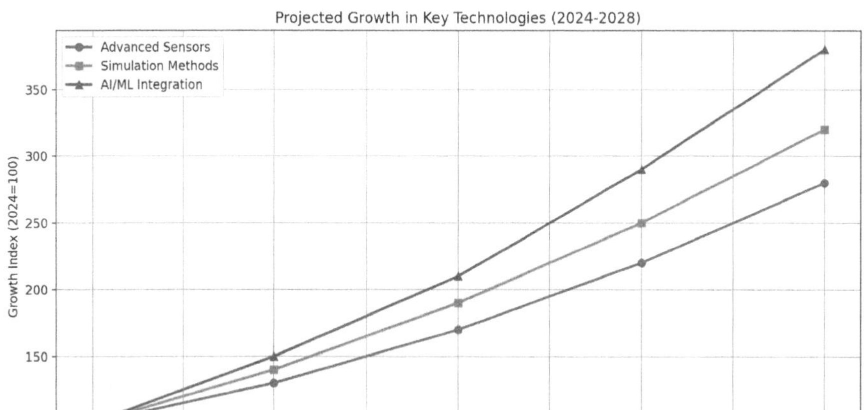

FIGURE 6.25 Projected growth in key technologies (2024–2028).

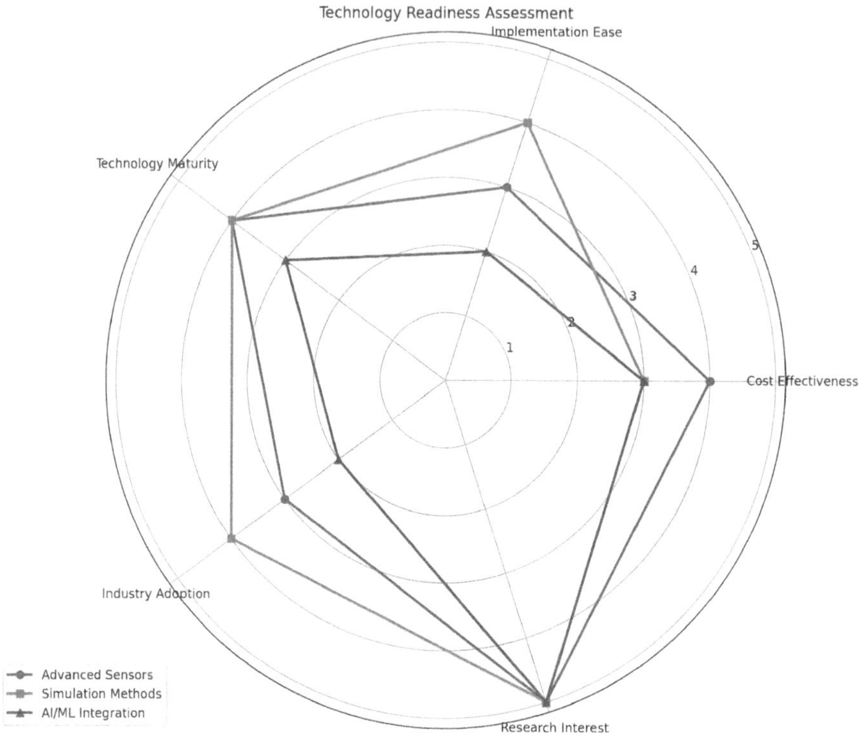

FIGURE 6.26 Technology readiness assessment.

However, among the subfields, AI/ML Integration has the highest research interest, indicating active research activity in this young and rapidly developing domain. This has given a comparison of the two technologies, showing the advantages and areas of improvement for each.

6.6 CONCLUSION

Thin-walled structures are an essential component of modern engineering due to their low weight and high strength-to-weight ratio. However, they face challenges such as buckling sensitivity, stress concentrations, and demanding material manufacturing processes. These are best solved through the use of computational tools to predict the structural performance of the structures, identify the best materials to use in the structures, and use superior fabrication methods to achieve the desired precision. It was found that the discrete use of simulation with a follow-up experimental validation of the designed structures is essential for the efficiency and adequacy of these structures. By using these strategies, engineers will be able to develop thin-walled structures that can satisfactorily meet the challenges posed by different applications. Multidisciplinary approach is of great importance in the design and analysis of thin-walled structures. By working together, material scientists, structural engineers, and manufacturers are part of the broader partnership that has led to stronger, more innovative solutions to secure building performance. The combination of these approaches provides the opportunity for structures to be more than an efficient structure; they become structures that can adapt to a wide variety of other applications and environmental conditions.

New testing methods, including non-destructive evaluation and real-time monitoring, will provide a deeper insight into the structural behavior. The development of comprehensive standards will help create consistency and reliability among various applications, leading to their wider use of thin-walled structures. Improved characterization techniques and standards will allow engineers to better understand the behavior of thin-walled structures and better develop more effective solutions for a variety of applications. On the one hand, thin-walled structures are highly efficient in terms of weight and usage of the material but tend to be affected by external pressure and are prone to buckling; on the other hand, thick-walled structures are more resistant to external pressures and less prone to buckling. In designing a screen, the final decision between thin- and thick-walled designs depends on your specific application requirements, such as the loads, the environmental conditions you are operating under, and of course, the cost. It is important to understand the tradeoffs that come with these two types of structures. Engineers can choose the best structure by carefully assessing the advantages and disadvantages of each type of structure available for their own purposes.

REFERENCES

1. Yang, L. I., G. A. O. Tong, Z. H. O. U. Qianying, C. H. E. N. Ping, Y. I. N. Dezheng, & Weihong Zhang. (2023). Layout design of thin-walled structures with lattices and stiffeners using multi-material topology optimization. *Chinese Journal of Aeronautics*, 36(4), 496–509.
2. Papadopoulos, V., & Papadrakakis, M. (2005) The effect of material and thickness variability on the buckling load of shells with random initial imperfections. *Computer Methods in Applied Mechanics and Engineering*, 194(12), 1405–426.
3. Chan, S. L., & Kitipornchai, S. (1987). Geometric nonlinear analysis of asymmetric thin-walled beam-columns. *Engineering Structures*, 9(4), 243–254.
4. Khosravi, P., Ganesan, R., & Sedaghati, R. (2008). Optimization of thin-walled structures with geometric nonlinearity for maximum critical buckling load using optimality criteria. *Thin-Walled Structures*, 46(12), 1319–1328.
5. Wysmulski, P., Falkowicz, K., & Filipek, P. (2021). Buckling state analysis of compressed composite plates with cut-out. *Composite Structures*, 274, 114345.
6. Bischoff, M., Bletzinger, K. U., Wall, W. A., & Ramm, E. (2004). Models and finite elements for thin-walled structures. In: Erwin Stein, René De Borst and Thomas Hughes (eds) *Encyclopedia of computational mechanics*. John Wiley & Sons, Ltd.
7. Zhang, S., Pan, Y., Li, N., & Sun, Q. (2014, February). A comparative study of the direct costs between prefabricated housing system and the traditional construction technology - A case study of precast concrete wall panel. In *Proceedings of the 18th International Symposium on Advancement of Construction Management and Real Estate*, Berlin, Heidelberg: Springer Berlin Heidelberg (pp. 49–358).

8. Perini, K., & Rosasco, P. (2013). Cost–benefit analysis for green façades and living wall systems. *Building and Environment*, 70, 110–121.
9. Capeto, A. P., Jesus, M., Uribe, B. E., Guimarães, A. S., & Oliveira, A. L. (2024). Building a greener future: advancing concrete production sustainability and the thermal properties of 3D-printed mortars. *Buildings*, 14(5), 1323.
10. Xie, S., Yang, W., Li, H., & Wang, N. (2017). Impact characteristics and crashworthiness of multi-cell, square, thin-walled, structures under axial loads. *International Journal of Crashworthiness*, 22(5), 503–517.
11. Drain, L. E. (1980). *The Laser Doppler Techniques.* John Wiley & Sons, Chichester.
12. Jenkins, C. H. & Tampi, M. (2000). Local membrane vibrations and inflatable space structures. In: *Proceedings of Space 2000*, ASCE (pp. 410–416).
13. Falkowicz, K. (2024). Stability and failure of thin-walled composite plate elements with asymmetric configurations. *Materials*, 17(9), 1943.
14. Castro, S. G., Zimmermann, R., Arbelo, M. A., Khakimova, R., Hilburger, M. W., & Degenhardt, R. (2014). Geometric imperfections and lower-bound methods used to calculate knock-down factors for axially compressed composite cylindrical shells. *Thin-Walled Structures*, 74, 118–132.
15. ASME. (2001). Alternative Rules for Determining Allowable External Pressure and Compressive Stresses for Cylinders, Cones, Sphere and Formed Heads, Section VIII, Divisions 1 and 2. In: *Cases of the ASME Boiler and Pressure Vessel Code, American Society of Mechanical Engineers*, New York, Standard No. Code Case 2286-2.
16. Wunderlich, W., & Albertin, U. (1998). Analysis and load carrying behaviour of imperfection sensitive shells." *Computational Mechanics: New Trends and Applications*, S. Idelsohn, E. Oñate, & E. Dvorkin, eds., International Centre for Numerical Methods in Engineering, Barcelona, Spain.
17. Krumm, M., Sauerwein, C., Hämmerle, V., & Heile, S. (2018, June). Robotic x-ray computed tomography system for fast automated non-destructive testing of joints in large assemblies and its application in automotive production. In: *12th European Conference on Non-Destructive Testing (ECNDT 2018) Proceedings, Gothenburg* (pp. 1–8).
18. Ifayefunmi, O., & Błachut, J. (2013). Instabilities in imperfect thick cones subjected to axial compression and external pressure. *Marine Structures*, 33, 297–307.
19. Raj, B., Jayakumar, T., & Rao, B. P. C. (1995). Non-destructive testing and evaluation for structural integrity. *Sadhana*, 20, 5–38.
20. David Müzel, S., Bonhin, E. P., Guimarães, N. M., & Guidi, E. S. (2020). Application of the finite element method in the analysis of composite materials: A review. *Polymers*, 12(4), 818.
21. Teng, J. G., & Rotter, J. M. (2004). "Buckling of thin shells." *Buckling of Thin Metal Shells*, J. G. Teng & J. M. Rotter, eds., Spon Press, London, pp. 1–41.
22. Godoy, L. A. (1996). *Thin-Walled Structures with Structural Imperfections.* Elsevier.
23. Casado, V. M., Hinsch, S., Garcia, J. G., & Castro, S. G. P. (2014). Effect of initial geometrical imperfection on the buckling load of cylindrical sandwich shells under axial compression. In: *13th European Conference on Spacecraft Structures, Material and Environmental Testing (ESA-SP)*, Braunschweig, Germany, April (pp. 1–10).
24. Lalin, V., Rybakov, V., & Sergey, A. (2014). The finite elements for design of frame of thin-walled beams. *Applied Mechanics and Materials*, 578, 858–863.
25. Weingarten, V. I., Seide, P., & Peterson, J. P. (1968). Buckling of thin-walled circular cylinders (No. NASA-SP-8007). Available at: https://ntrs.nasa.gov/api/citations/20205011530/downloads/20205011530%20Rev%202FINALa%201-2023.pdf (Accessed on: 19/12/2025)

26. Wankhede, S., Pesode, P., Pawar, S., & Lobo, R. (2023). Comparison study of GRA, COPRAS and MOORA for ranking of phase change material for cooling system. In: Materials Today: Proceedings, Hyderabad, India.
27. Pesode, P., Barve, S., Wankhede, S. V., Pawar, S. K., & Jadhav, D. R. (2023, December). Antibacterial activities of plasma electrolytic oxidation coated magnesium alloys. In: *2nd International Conference & Exposition on Mechanical, Material, and Manufacturing Technology,* Hyderabad, India, *AIP Conference Proceedings* 2943, 030007.
28. Zabler, S., Maisl, M., Hornberger, P., Hiller, J., Fella, C., & Hanke, R. (2021). X-ray imaging and computed tomography for engineering applications. *tm-Technisches Messen*, 88(4), 211–226.
29. Romanoff, J., Körgesaar, M., & Remes, H. (2020). Emerging challenges for numerical simulations of quasi-static collision experiments on laser-welded thin-walled steel structures. *Journal of Marine Science and Application*, 19(4), 567–583.
30. Vurtur Badarinath, P., Chierichetti, M., & Davoudi Kakhki, F. (2021). A machine learning approach as a surrogate for a finite element analysis: Status of research and application to one dimensional systems. *Sensors*, 21(5), 1654.
31. Bathurst, R. J., Allen, T. M., & Walters, D. L. (2002). Short-term strain and deformation behavior of geosynthetic walls at working stress conditions. *Geosynthetics International*, 9(5–6), 451–482.
32. Alfrey, G. F. (1980). The laser Doppler technique. *Physics Bulletin*, 31(9), 319.
33. Buligan, G., Paone, N., Revel, G. M. & Tomasini, E. P. (2003). The measurement of hermetic compressor valve motion by laser vibrometry. Senior N&V Researcher Via Consorziale. 13.
34. Bachalo, W. D., & Houser, M. J. (1984). Phase/Doppler spray analyzer for simultaneous measurements of drop size and velocity distributions. *Optical Engineering*, 23(5), 235583.
35. Bloss, B., & Rao, M. D. (2002). Measurement of damping in structures by the power input method. *Experimental Techniques*, 26(3), 30–32.
36. Zappino, E., Cavallo, T., & Carrera, E. (2016). Free vibration analysis of reinforced thin-walled plates and shells through various finite element models. *Mechanics of Advanced Materials and Structures*, 23(9), 1005–1018.
37. Tabatabai, H., Oliver, D. E., Rohrbaugh, J. W., & Papadopoulos, C. (2013). Novel applications of laser Doppler vibration measurements to medical imaging. *Sensing and Imaging: An International Journal*, 14, 13–28.
38. Ratner, A., Wood, M., Chowanietz, M., Kumar, N., Patel, R., Hadlum, P., ... & Masters, I. (2022). Laser doppler vibrometry for evaluating the quality of welds in Lithium-Ion supercells. *Energies*, 15(12), 4379.
39. Shiraiwa, T., Ishikawa, K., Enoki, M., Shinozaki, I., & Kanazawa, S. (2020). Acoustic emission analysis using Bayesian model selection for damage characterization in ceramic matrix composites. *Journal of the European Ceramic Society*, 40(8), 2791–2800.
40. Proctor, T. M. (1982). An improved piezoelectric acoustic emission transducer. *The Journal of the Acoustical Society of America*, 71(5), 1163–1168.
41. Glaser, S. D., Weiss, G. G., & Johnson, L. R. (1998). Body waves recorded inside an elastic half-space by an embedded, wideband velocity sensor. *The Journal of the Acoustical Society of America*, 104(3), 1404–1412.
42. Tulshibagwale, N., Brodnik, N., Muir, C., Hilmas, A., Kiser, J. D., Smith, C. E., ... & Daly, S. (2024). Acoustic emission in ceramic matrix composites. *Applied Mechanics Reviews*, 76(4), 040801.

43. Kastner, J., Heinzl, C., Plank, B., Salaberger, D., Gusenbauer, C., & Senck, S. (2017, February). New X-ray computed tomography methods for research and industry. In: *7th Conference on Industrial Computed Tomography*, Leuven, Belgium, Leuven (pp. 1–10).
44. Civera, M., & Surace, C. (2022). Non-destructive techniques for the condition and structural health monitoring of wind turbines: A literature review of the last 20 years. *Sensors*, 22(4), 1627.
45. Doyle, J. F. (2013).Thin Plates and Shells, In: James F. Doyle (Author) *Nonlinear analysis of thin-walled structures: statics, dynamics, and stability*, Springer Science & Business Media, Springer New York, NY (pp. 90–168).
46. Donnelly, N., Bibas, A., Jiang, D., Bamiou, D. E., Santulli, C., Jeronimidis, G., & O'Connor, A. F. (2009). Effect of cochlear implant electrode insertion on middle-ear function as measured by intra-operative laser Doppler vibrometry. *The Journal of Laryngology & Otology*, 123(7), 723–729.
47. Addessi, D., Di Re, P., & Cimarello, G. (2021). Enriched beam finite element models with torsion and shear warping for the analysis of thin-walled structures. *Thin-Walled Structures*, 159, 107259.
48. Ye, J., Quan, G., Kyvelou, P., Teh, L., & Gardner, L. (2022, April). A practical numerical model for thin-walled steel connections and built-up members, *Structures*, 38, 753–764.
49. Rasmussen, K. J. R. (2000). General report–experimental techniques in the testing of thin-walled structural members. In: *Coupled Instabilities in Metal Structures: CIMS'2000*, Lisbon, Portugal (pp. 225–239).
50. Tang, D., Yang, C., Huang, Y., & Ku, D. N. (1999). Wall stress and strain analysis using a three-dimensional thick-wall model with fluid–structure interactions for blood flow in carotid arteries with stenoses. *Computers & Structures*, 72(1–3), 341–356.
51. Liu, W., & Ewing, M. S. (2007). Experimental and analytical estimation of loss factors by the power input method. *AIAA Journal*, 45(2), 477–484.
52. Woodhams, F. (1980). The laser Doppler technique. *Optica Acta: International Journal of Optics*, 27(11), 1501–1501.
53. Rothberg, S. J., Allen, M. S., Castellini, P., Di Maio, D., Dirckx, J. J. J., Ewins, D. J., ... & Vignola, J. F. (2017). An international review of laser Doppler vibrometry: Making light work of vibration measurement. *Optics and Lasers in Engineering*, 99, 11–22.
54. Rosso, A., Almeida, J. P., & Beyer, K. (2016). Stability of thin reinforced concrete walls under cyclic loads: State-of-the-art and new experimental findings. *Bulletin of Earthquake Engineering*, 14, 455–484.
55. Wu, W. X., Shu, C., & Wang, C. M. (2007). Vibration analysis of arbitrarily shaped membranes using local radial basis function-based differential quadrature method. *Journal of Sound and Vibration*, 306(1–2), 252–270.
56. Kabir, M., Kazari, H., & Ozevin, D. (2018). Piezoelectric MEMS acoustic emission sensors. *Sensors and Actuators A: Physical*, 279, 53–64.
57. Benedict, H. E., & Reddy, J. N. (2021). Review of finite element modeling and analysis applications in aerospace structural systems design: Design conceptualization tool. *International Journal of Aerospace and Mechanical Engineering*, 8(2), 10–16.
58. Liu, F., Chen, M., Liu, S., Xiang, Z., Huang, S., Lim, E. G., & Zhang, S. (2024). Stress-driven generative design and numerical assessment of customized additive manufactured lattice structures. *Materials & Design*, 241, 112956.
59. Wankhede, S., Lobo, R., & Pesode, P. (2024). Evaluating machine learning algorithm for real-time heat exchanger optimization and automatic issue detection device: Experimental analysis. *International Journal on Interactive Design and Manufacturing (IJIDeM)*, 18(7), 4409–4420.

60. Jia, H., et al. (2020). Finite element analysis and simulation of materials. *Materials*, 13(10), 2200.
61. Nath, D., Neog, D. R., & Gautam, S. S. (2024). Application of machine learning and deep learning in finite element analysis: A comprehensive review. *Archives of Computational Methods in Engineering* 31(5), 2945–2984.
62. Shi, Bastian. (2021). *Abaqus vs.* SolidWorks: Dawn of FEA.
63. Wang, Z., Bhaduri, A., Xu, H., & Wang, L. (2025). An uncertainty-aware deep learning framework-based robust design optimization of metamaterial units. *Structural and Multidisciplinary Optimization*, 68(3), 1–26.
64. Chaabani, H., Mesmoudi, S., Boutahar, L., & Bikri, K. E. (2022). Buckling of porous FG sandwich plates subjected to various non-uniform compressions and resting on Winkler–Pasternak elastic foundation using a finite element model based on the high-order shear deformation theory. *Acta Mechanica*, 233(12), 5359–5376.
65. Feng, M., Wang, Y. C., & Davies, J. M. (2003). Structural behaviour of cold-formed thin-walled short steel channel columns at elevated temperatures. Part 2: Design calculations and numerical analysis. *Thin-Walled Structures*, 41(6), 571–594.
66. Bull, J. W. (2002). *Finite Element Applications to Thin-Walled Structures.* CRC Press.
67. Jolly, M. R., Prabhakar, A., Sturzu, B., Hollstein, K., Singh, R., Thomas, S., ... & Shaw, A. (2015). Review of non-destructive testing (NDT) techniques and their applicability to thick walled composites. *Procedia CIRP*, 38, 129–136.
68. Al-Bermani, F. G., & Kitipornchai, S. (1990). Nonlinear analysis of thin-walled structures using least element/member. *Journal of Structural Engineering*, 116(1), 215–234.
69. Matias, M., Almeida, F., Moura, R., & Barraca, N. (2021). High resolution NDT in the characterization of the inner structure and materials of heritage buildings walls and columns. *Construction and Building Materials*, 267, 121726.
70. Yukawa, K., Cohen, L., Blamey, P., Pyman, B., Tungvachirakul, V., & O'Leary, S. (2004). Effects of insertion depth of cochlear implant electrodes upon speech perception. *Audiology and Neurotology*, 9(3), 163–172.
71. Li, Y., Zhou, Z., & Wang, J. (2019). Analysis of linear non-destructive testing and evaluation methods for thin-walled structure inspection using ultrasonic array. *Coatings*, 9(2), 146.
72. Liew, J. R., Thevendran, V., & Shanmugam, N. E. (Eds.). (1998). *Thin-Walled Structures: Research and Development.* Elsevier.
73. Loughlan, J. (2018). Section XIII: Thin-Walled Beams and Columns. In: J. Loughlan (eds), *Thin-walled structures: Advances in research, design and manufacturing technology.* CRC Press LLC.
74. Halis, S., & Altın, M. (2024). Numerical investigation of energy absorption performance in thin-walled structure under three-point bending test. *International Journal of Automotive Science and Technology*, 8(1), 159–166.
75. Kazaz, İ., Gülkan, P., & Yakut, A. (2012). Performance limits for structural walls: An analytical perspective. *Engineering Structures*, 43, 105–119.
76. Camotim, D., & Basaglia, C. (2014). On the behaviour, failure and direct strength design of thin-walled steel structural systems. *Thin-Walled Structures*, 81, 50–66.
77. Abrams, M. S., D'Souza, M. V., Higginson, P., Metes, W. S., Quintiere, J. G., Parker, W. J., ... & White Jr, J. A. (1985). Review of ASTM test for surface burning characteristics of building materials (E 84)(an ASTM White paper). *Journal of Testing and Evaluation*, 13(2), 89–114.

78. Camotim, D., Silvestre, N., Gonçalves, R., & Dinis, P. B. (2006). GBT-based structural analysis of thin-walled members: Overview, recent progress and future developments. *Advances in Engineering Structures, Mechanics & Construction*, Part of the book series: Solid Mechanics and Its Applications (SMIA, volume 140), 187–204.
79. Papangelis, J. P., & Hancock, G. J. (1995). Computer analysis of thin-walled structural members. *Computers & Structures*, 56(1), 157–176.
80. Mahajan, G., & Narkhede, P. (2024). Integrating BIM with Digital Technology Trends in the Construction Industry: Implementation Insights for 2023. *Library of Progress-Library Science, Information Technology & Computer*, 44(3), 20283–20308.
81. Balbaud, F. et al., (2021). A NEA review on innovative structural materials solutions, including advanced manufacturing processes for nuclear applications based on technology readiness assessment. *Nuclear Materials and Energy*, 27, 101006.

7 Post-Processing on Thin-Walled Structures

K. G. Anbarasu, M. Satthiyaraju, K. G. Ashok, and R. Majusha

7.1 INTRODUCTION

Structures made of high-density materials result in higher fuel consumption, increased weight, and higher emissions. Recently, many countries made an agreement to reduce global warming by <2°C. An increase in global warming results in a rise in temperature and leads to negative impacts on the environment [1]. However, from the perspective of launch vehicles, the reduction in weight will help to increase the number of satellites in each launch. Another major problem worldwide is that more than 1.2 million people die annually in traffic accidents [2]. Furthermore, the engineering components, such as turbine blades, rocket engine casings, nozzles, and combustion chambers, need cooling channels for better functional performance and life. The above-mentioned problems, such as increased fuel consumption, catastrophic failure of automobile parts during accidents, and a higher rate of heat transfer, can be resolved by using thin-walled structures with a high strength-to-weight ratio and energy absorption (EA) characteristics. This chapter elaborates on the method of making thin-walled structures in Section 7.2.

In recent times, additive manufacturing processes, or 3D printing processes, are used to make thin-walled structures due to their ability to make lightweight structures at low cost and in less time, with more design freedom than conventional processes. Furthermore, the conventional processes pose difficulty in making complex and customised features such as turbine blades with internal cooling features, rocket engines with regenerative cooling features, rocket nozzles with cooling features, lightweight mirrors, heat exchangers with lattice structures, and so on [3–6]. The other problems associated with conventional methods are wastage of materials, being labour-intensive, a greater number of assemblies and subcomponents for making a product, and the requirement of special tools and dies. However, additive manufacturing poses challenges such as poor surface quality, anisotropic properties, inhomogeneous microstructure, dimensional inaccuracy, form inaccuracy, residual stresses, and partially melted as well as loose powders on the internal features. The in-depth details about each challenge that occurred during the additive manufacturing of thin-walled structures are elaborated in Section 7.3.

DOI: 10.1201/9781003571209-7

To overcome these challenges, further post-processing methods such as machining, heat treatment, hot isostatic pressing (HIP), peening, polishing, and blasting were carried out in industries. More details about the post-processing methods used for additive manufactured (AM) thin-walled structures are given in Section 7.4. Moreover, the changes obtained after each post-processing method are elaborated in subsequent Sections 7.4.1 and 7.4.2. As aforementioned, conventional methods consume more time and money for making parts like turbine blades, bio-implants, electronic cooling systems, combustion chambers, thrust chambers, nozzles, mirrors, and so on. Also, the conventional process is labour-intensive, and more material wastage occurs. This induces aerospace, space, automobile, biomedical, and electronics system manufacturers to use the additive manufacturing processes for making products with design freedom to maximise efficiency and achieve better functional performance with a longer life. The recent application of AM thin-walled structures in various domains is given in Section 7.5.

As aforementioned, though additive manufacturing processes have a lot of advantages, they also have demerits. Owing to this, post-processing steps were carried out for AM parts. However, the post-processing steps add cost and time to making the parts. Therefore, reducing the post-processing time and cost should be a key consideration. Simultaneously, the quality of the product should not be compromised. For example, powders that are trapped inside the internal features reduce the overall performance of the product. Likewise, the presence of porosity causes leakages, and the presence of micro-irregularities, as well as form error, affects the functional performance of thin-walled structures [4,5]. In recent times, process monitoring was carried out to understand the process physics using sensors and cameras. With the sensor data and camera data, artificial intelligence models were developed for predicting the surface quality, defects, and mechanical properties [7,8]. Also, it helps for inspection purposes and planning the next processing steps. For example, predicting the defect depth helps the manufacturers to plan for subsequent processes like polishing. One of the recent trends in the scientific community is manufacturing in space. Among various research activities of manufacturing in space programmes, the research on additive manufacturing of thin-walled structures in microgravity and zero gravity is one of the recent trends. More details about the future trends in post-processing of thin-walled structures are given in Section 7.5.

7.2 DIFFERENT METHODS USED FOR MAKING THIN-WALLED STRUCTURES

The commonly used methods to make thin-walled structures are given below:

I. **Extrusion** – in this process, the thin-walled structure is produced by forcing material through a shaped die for manufacturing components such as tubes, channels, and frames (Figure 7.1(a)) [9]. This is mainly used for its consistent wall thickness and efficient mass production.
II. **Machining** – machining is used to finish thin-walled structures by accurately removing material using cutting tools. The milling process is used for machining thin-walled structures (Figure 7.1(b)) [10]. It is particularly useful for obtaining accurate dimensions and surface quality in lightweight components.

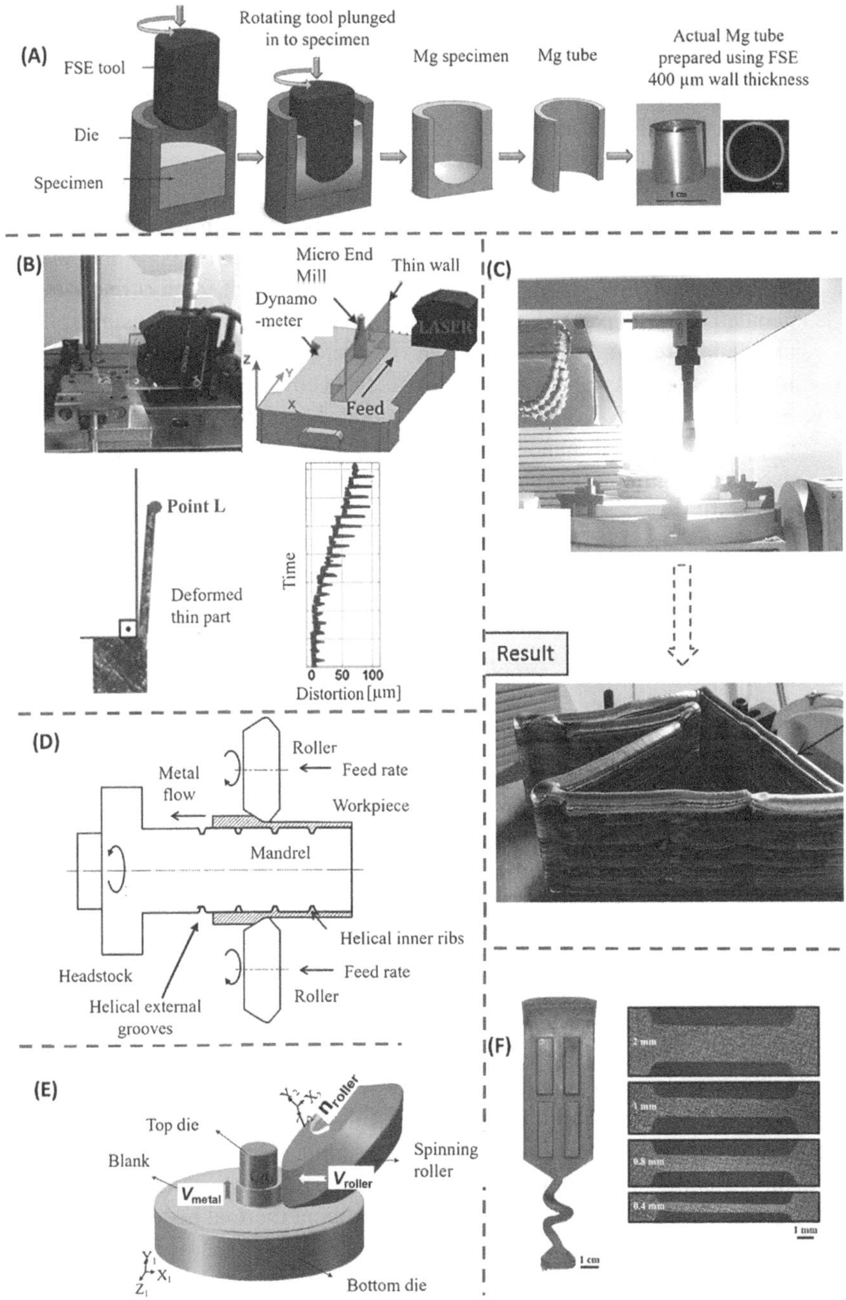

FIGURE 7.1 Different manufacturing methods used for making thin-walled structures: (a) extrusion [9], (b) machining [10], (c) additive manufacturing [11], (d) forming [12], (e) spinning [13], and (f) casting [14].

III. **Additive Manufacturing** – the additive manufacturing process produces thin-walled structures directly from the digital model. It is a very flexible process. The thin-walled part is produced by layer by layer of material deposition or fusion of powders (Figure 7.1(c)) [11]. Nowadays, the parts produced using this process are applied in aerospace, medical, and energy applications due to their ability to produce complex shapes with design freedom.
IV. **Sheet Metal Forming** – is widely used to fabricate thin-walled structures by plastically deforming flat metal sheets into various shapes without material removal. This process involves dies and a punch for forming simple geometries, whereas cylindrical thin-wall tubes are formed using rollers and a mandrel (Figure 7.1(d)) [12]. It is a preferred technique for producing lightweight panels, enclosures, and structural covers.
V. **Spinning** – it involves shaping a thin metal sheet into a desired form using a spinning tool and mandrel (Figure 7.1(e)) [13]. This method is commonly used for manufacturing cones, domes, and cylindrical shells with uniform thin walls.
VI. **Casting** – is highly useful for creation of complex thin-walled structures by pouring molten material into a mould where it solidifies into the required geometry (Figure 7.1(f)) [14].

This method is suitable for complicated shape components such as housings and casings with thin sections. The most commonly used casting process for making thin-walled structures is investment casting.

7.3 CHALLENGES IN USING ADDITIVE MANUFACTURED THIN-WALLED STRUCTURES

The AM thin-walled structures cannot be directly used for certain applications, such as aircraft structures, rocket engines, mirrors, thrust chambers, combustion chamber liners, heat exchangers, bio-implants, and electronic cooling systems, due to the stringent requirements of standards [3–6]. In general, as-built (AB) AM thin-walled structures have poor surface quality, residual stress, residual powders in the internal features, part distortion, warpage, cracks, anisotropic properties, and dimensional deviations. Heat accumulation during printing causes changes in melt pool characteristics, which in turn results in dimensional inaccuracies [4]. Likewise, the heat accumulation increases the size of the top layer more than that of the bottom layer due to the widening of the melt pool, which results in geometrical inaccuracy or form error [4]. Moreover, the AM part experiences different thermal histories and eventually leads to an inhomogeneous microstructure as well as anisotropic properties. The small grain size results in a higher number of grain boundaries. These changes will directly affect the heat transfer characteristics [6]. One of the crucial problems in the AB AM part is residual stress. This occurs due to the repeated melting and solidification during the layer-by-layer manufacturing process [4].

The repeated melting and solidification cause shrinkage. The shrinkage is constrained by the previously deposited layer [4]. This induces tensile residual stresses on the newly deposited layer. Because of this, metal part distortion occurs on the thin-walled structure. Generally, the top and bottom regions of AM thin-walled structures have tensile residual stresses, and the middle region experiences both tensile and compressive residual stresses. Owing to these residual stresses, cracks occur on the AM part [4]. In addition to this, the presence of tensile residual stress has a negative impact on the life of the part. Besides, the presence of defects such as pores and voids results in fluid leaks. Defects like pores and voids not only create leakage issues but also affect the load-bearing characteristics and cause conduction barriers [3,6]. The phenomena, such as lack of fusion, keyholing, and trapped gases, create pores and voids. The lack of fusion occurs due to insufficient beam power or at higher scanning speeds, and the keyholing occurs at high-energy beam conditions. The porosity caused by trapped gases in the powders occurs when these gases are released during melting and become locked in during solidification. Similar to residual stress and porosity, the other vital challenge concerning AB AM thin-walled structures is their poor surface quality [4].

A typical AB metal AM surface contains loose powders, partially melted powders, balling effects, layer-by-layer manufacturing marks, porosities, and semi-welded tracks [15]. As a result, the AB part exhibits poor surface quality. The surface roughness has a significant impact on the fatigue life of the thin-walled structure since the peaks act as crack initiation sites due to stress concentration. Furthermore, the high surface roughness in the thin-walled mirrors affects the functional performance, such as reflectivity. In fluid flow applications, the presence of a rough surface on the thin-walled structure creates pressure drop and turbulence [4,6]. On the other side, the presence of a rough surface increases the dimension of the thin-walled structure's external surface and reduces the dimension of the internal surface, like a hole or pipe [6]. After finishing, either the dimension increases or decreases depending on the interior or exterior surface. In particular, the powders that adhere to the circumference of the internal features cause blockages and also increase the roughness of those features [6].

This blockage directly affects the functional performance and life of the part, like dies with conformal cooling features, turbine blades with internal cooling features, bio-implants, etc. For example, the residual powders or partially melted material present inside the cooling features may melt during the operation. In the case of a bio-implant, the presence of powder causes inflammation and blocks the blood vessels [16]. An inclined or overhanging feature is commonly present in mechanical components like turbine blades, impellers, and so on. Such products need support structures during printing to prevent parts from warping, distortion, and heat dissipation. However, removing the support structure needs extra post-processing, and, in some cases, removing the support structure is challenging due to intricate features [4]. Furthermore, one of the required characteristics for using an AB AM thin-walled structure in saline water, acidic, alkaline, and human body environmental conditions is its corrosion resistance. Typically, a rough surface with surface defects results in a higher corrosion rate. Therefore, to functionalise the AB AM thin-walled parts, further post-treatments are inevitable. The post-treatments used for functionalising the AB AM thin-walled structure are elaborated in the subsequent sections.

7.4 POST-PROCESSING METHODS USED FOR THIN-WALLED STRUCTURES

To use AM thin-walled structures for various applications, the printed thin walls must meet the following requirements:

- Good surface integrity
- Better dimensional and geometrical accuracy
- Reduced porosity and increased density
- Refined microstructure
- Improved mechanical properties

The above requirements can be achieved through post-processing. In general, post-processing is classified into two major categories: (1) primary post-processing and (2) secondary post-processing.

Primary post-processing primarily focuses on the removal of support structures, removal of powders, and cleaning of the printed part. Generally, the removal of the support structure is done by using conventional machining, wire electric discharge machining, and the sawing process. The unfused or residual powders are removed by a vacuum cleaning process and an ultrasonic cleaning process [6]. On the other hand, secondary post-processing aims to enhance surface integrity, improve geometrical and dimensional accuracy, refine the microstructure, and enhance mechanical properties. Consequently, secondary post-processing has a significant impact on the functional performance and life of thin-walled structures.

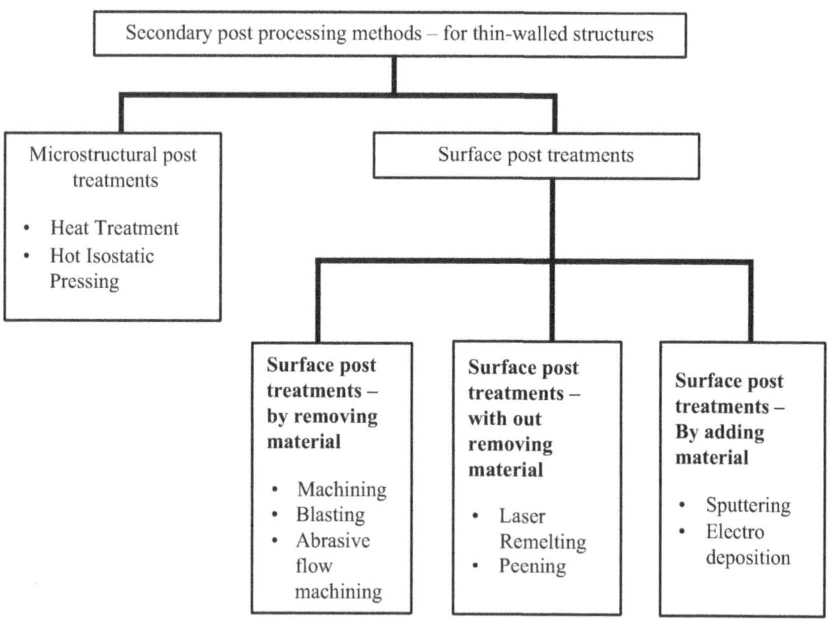

FIGURE 7.2 Classification of secondary post-processing methods used for thin-walled structures.

Secondary post-processing is usually done in two ways: (1) to improve the microstructure and its related properties, and (2) to make the surface better and improve size and shape accuracy. The broad classification of secondary post-processing methods used for thin-walled structures is depicted in Figure 7.2. In Figure 7.2, the classification is made based on the purpose of post-treatments and the way followed for surface modification. Details about each method are elaborated in the following sections.

7.4.1 Microstructural Post-Treatments

Microstructural post-treatments are performed to modify the microstructure and achieve uniformity along both the vertical and horizontal directions of printing. These modifications result in improved properties such as strength, ductility, hardness, density, and EA capacity, among others. Additionally, internal stresses caused by localised melting and rapid solidification are relieved during microstructural post-treatments. The most commonly used microstructural post-treatments are heat treatment and HIP. Details about each method are elaborated in the following sections.

7.4.1.1 Effect of Heat Treatment

In general, heat treatment processes are used to relieve internal stresses and improve properties such as toughness, ductility, strength, hardness, and more. Additionally, heat treatment is employed for microstructural refinement. The heat treatment processes used for AM thin-walled structures are listed below, and their effects on density, porosity, mechanical properties, and microstructure are elaborated in the subsequent sections:

- Solution heat treatment (SHT)
- Aging
- Stress-relieving heat treatment
- Annealing

One of the major factors affecting the strength and stability of thin-walled structures is their porosity and density. The density of AB Selective Laser Melting (SLM)-printed Al-Mg-Si alloy thin-walled structures exhibits varying relative density (RD) with changes in wall thickness (0.5–5 mm). Similarly, AB parts after SHT and artificial aging (AA) follow the same trend in RD. For all cases (AB, SHT, and AA), the maximum RD is obtained for a 0.5 mm thick wall. Thinner walls have fewer pores, resulting in higher RD [17]. The observed porosity is attributed to inadequate melting and weak interlayer bonding of powder particles [17]. Among all heat treatment conditions, the thin wall (0.5 mm) treated by AA produces the highest RD. This is due to the formation of needle-like precipitates of Mg_2Si during the aging process. However, the AA treatment is less effective for higher wall thicknesses [17]. The SHT and AA-treated 0.5 mm thick part shows a reduction in pore size compared to the AB condition. However, for higher wall thicknesses (1.5 mm), the micropores merge during the SHT treatment, resulting in larger pore sizes, whereas the AA-treated parts exhibit smaller pores [17].

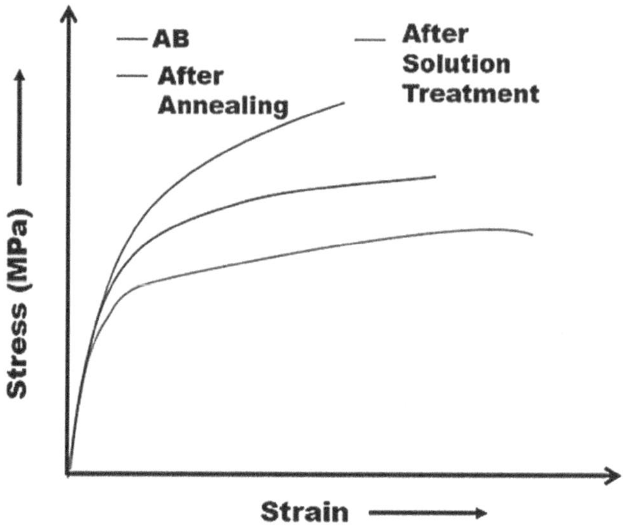

FIGURE 7.3 Typical engineering stress–strain curves of SLM-printed AlSi10Mg parts under different heat treatment conditions.

Also, heat treatment has a significant impact on mechanical properties such as ductility, hardness, tensile strength, yield strength, and fatigue strength. Additionally, one of the ways to relieve stress is through post-heat treatment [2]. The AB SLM -printed AlSi10Mg alloy thin-walled parts exhibit higher tensile and yield strengths of about 300 MPa and 220 MPa, respectively. However, the ductility is relatively low (4%). After post-heat treatments such as annealing (300°C for 2 hours) and SHT (530°C for 2 hours), the ductility improves by 6% and 11%, respectively.

On the other hand, from Figure 7.3, it was observed that the tensile and yield strength decrease after both treatments (annealing and SHT) [2]. The observed improvement in ductility is due to microstructural coarsening. The AB thin-walled part has a primary Al matrix with a white fibrous Si network, resulting in higher strength than the heat-treated conditions. Post-heat treatment results in microstructure coarsening, shown in Figure 7.4 [2]. Additionally, varying the heat treatment (annealing) temperature from 260°C to 440°C for 1 hour, with a step size of 30°C, shows differences in ductility and strength [18]. The maximum tensile and yield strengths are obtained in the temperature range of 260°C to 290°C, while the highest ductility is achieved at 440°C [18]. However, from Figure 7.5, in the case of Wire Arc Additive Manufactured (WAAM) thin-walled parts made of 18Ni 300 maraging steel, lower ultimate tensile strength (UTS) was observed in both horizontal (1320 MPa) and vertical (1316 MPa) directions compared to parts after direct aging (AT) and solution + aging (SAT) [19]. The UTS obtained after AT and SAT in both directions are 1744 MPa (horizontal), 1693 MPa (vertical), 1605 MPa (horizontal), and 1565 MPa (vertical), respectively.

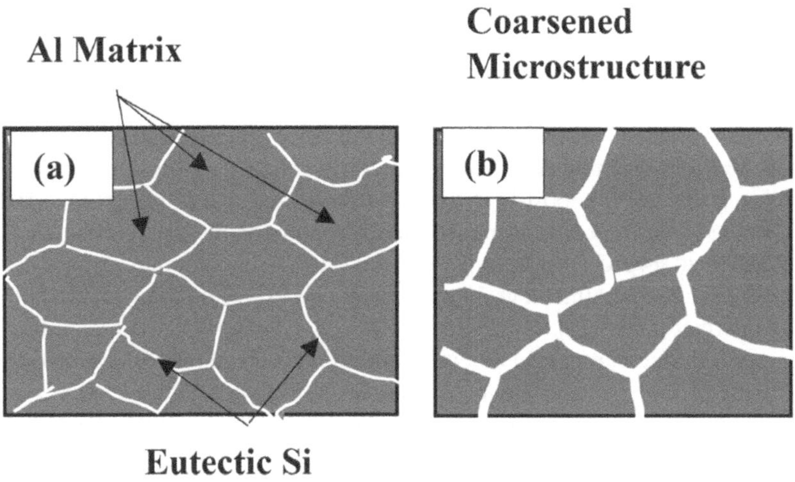

FIGURE 7.4 Typical microstructure of SLM-printed AlSi10Mg parts (a) as-built and (b) after heat treatment.

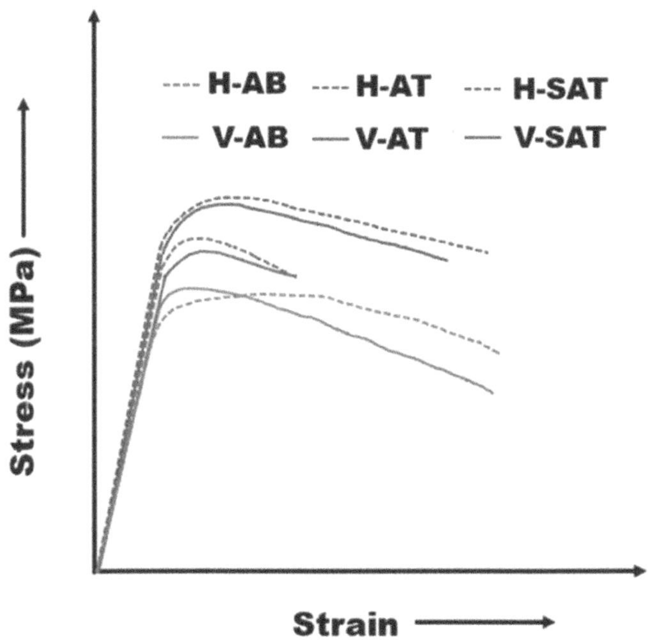

FIGURE 7.5 Typical engineering stress–strain curve of WAAM-printed 18Ni 300 maraging steel part under different heat treatment conditions.

Also, the ductility of the AB thin-walled part was higher in both directions compared to the post-heat-treated parts [19]. The observed higher UTS for the post-heat-treated samples is due to the formation of nano-precipitates [19]. Overall, the AT-treated part shows higher UTS than the SAT-treated part, but the SAT-treated parts exhibit less anisotropy [19]. Additionally, the post-heat-treated thin-walled parts have higher EA capability compared to the AB thin-walled parts. The EA values for AB, annealed, and SHT thin-walled parts are 93.0 KJ, 232.5 KJ, and 315.8 KJ, respectively. Moreover, the specific EA of the SHT-treated thin-walled part is 148% higher than that of the AB thin-walled part [2]. Furthermore, the microstructure of the post-heat-treated thin-walled parts transforms from a non-homogeneous to a homogeneous microstructure as the temperature increases from 290°C to 440°C [18]. Initially, the AB part has a columnar grain structure, but after post-heat treatment, grain refinement occurs, resulting in a more homogeneous microstructure [18].

In general, due to repeated and rapid thermal cycles in the SLM additive manufacturing process, a heterogeneous microstructure with a high level of residual stress and elemental segregation is formed. In order to relieve the stress and obtain a homogeneous microstructure, stress-relieving post-heat treatment processes are carried out. The stress-relieving heat treatment (800°C and 870°C for 1 hour) results in more equiaxed grains, whereas the SLM-printed IN625 thin-walled structures without any treatment have long columnar structures along the build direction [20]. However, after stress-relieving post-heat treatment (1100°C for 1 hour) on laser powder – directed energy deposition (LP-DED) C103 thin-walled parts, strength is reduced, and ductility is increased in both directions [21]. This occurs due to the reduction in residual stress by the diffusion of dislocations to the grain boundaries.

Another important property for critical applications with cyclic loads is fatigue strength. The post-heat-treated (T6 Quenching) SLM-printed AlSi10Mg thin-walled part shows an 80% improvement in fatigue strength [22]. Likewise, in some applications, wear resistance plays a crucial role. Hardness is one of the primary properties required to achieve high wear resistance characteristics. Generally, heat treatment results in higher hardness, and similar behaviour was observed in the case of post-heat-treated AM thin-walled structures [19].

7.4.1.2 Effect of Hot Isostatic Pressing

Generally, the HIP process is used to increase the density and strength, as well as to reduce the porosity of AM parts. The AB LP-DED C103 thin-wall parts show a 7% increase in strength and a 27% increase in ductility after HIP post-treatment (1560°C and 150 MPa for 3 hours). Moreover, from Figure 7.6, it was observed that the HIP treatment increases the UTS by 7.68% and the yield strength by 3.69% [21]. The microstructure of the HIP (920°C, 1000 bar, 2 hours) treated AB EBM TiAl4V alloy thin-walled part has a homogeneous microstructure in all cross-sectional planes (shown in Figure 7.7). Owing to this homogeneous microstructure, the mechanical properties of the HIP-treated parts are more uniform than the AB condition [23]. However, an increase in dimension was found after HIP treatment of the combustion chamber [24].

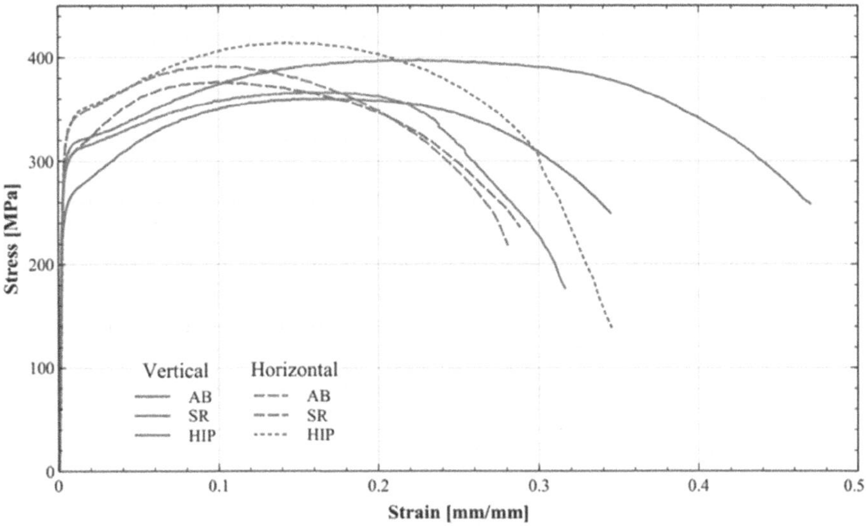

FIGURE 7.6 Stress–strain curves in vertical and horizontal directions for the LP-DED C103 thin-wall part treated under various post-treatments [21].

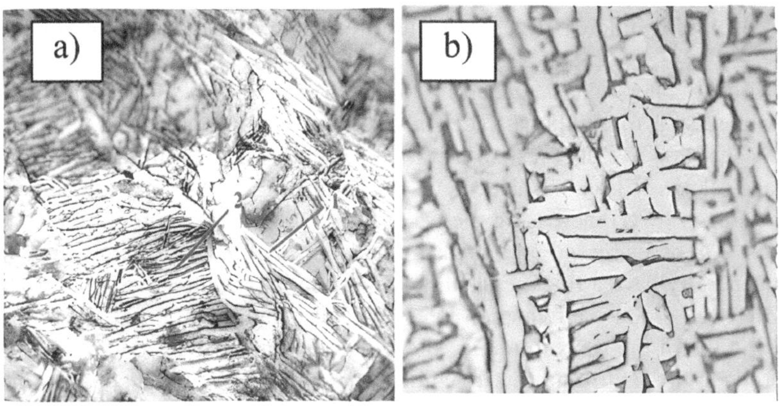

FIGURE 7.7 Microstructure of EBM TiAl4V alloy (A) before HIP (B) after HIP [23]. HIP, Hot isostatic pressing.

7.4.2 Surface Post-Treatments

Surface integrity has a significant impact on the functional performance of thin-walled structures used in aircraft and automobiles. The fatigue strength, hardness, fluid flow characteristics, and wear resistance properties of the thin-walled structures are influenced by surface integrity. Additionally, for assembly purposes, the geometrical and dimensional accuracy of the thin-walled parts plays a crucial role. In general, AM parts have poor surface integrity, meaning that AB AM thin-walled parts do not meet stringent requirements. To meet these requirements, further post-processing is carried out.

To improve dimensional and geometrical accuracy, as well as to enhance surface finish, surface post-treatments such as machining are performed. Furthermore, surface post-treatments are carried out to improve surface hardness and surface finish, and to reduce porosity as well as residual stress. Surface treatments without material removal or transformative processes, such as laser remelting (LR), laser shock peening (LSP), shot peening, and ultrasonic peening, are carried out. For some applications, the surface should be corrosion-resistant, oxidation-resistant, and bacteria-resistant. Generally, coatings are used to achieve these properties. The post-surface treatments used for thin-walled structures are elaborated in the subsequent sections.

7.4.2.1 Surface Modification by Material Removal

The AB AM thin-walled parts have a rough finish, and their geometrical and form accuracy is low. Generally, machining processes are used to improve the surface finish and dimensional as well as form accuracy. However, machining thin-walled AM is challenging due to the following reasons:

- The anisotropic nature of AM parts results in machining instability
- Machining of low-stiffness structures results in deflection and vibration

The milling operation is generally used for machining. When milling AM thin-walled structures, the surface finish and accuracy depend on wall stiffness [25]. Machining low-stiffness thin-walled structures results in vibration (chattering) and relatively large displacements of the milling tool and part. This problem can be solved by (1) increasing the wall stiffness and (2) reducing the MRR. The stiffness can be increased by increasing the wall thickness, but thin-walled structures generally have a thickness limitation (<5 mm). On the other hand, reducing the MRR can produce a reduction in cutting force, but it will affect the productivity [25]. One of the ways to reduce the deflection and chatter, as well as increase the productivity, is to find the dynamic behaviour (frequency response function) of the thin-walled structure during milling [25]. A finite element method model is used to predict the frequency response function. Based on the aspect of obtaining good surface finish, the adopted machining cycle with three phases (roughing, semi-finishing, and finishing) was used. The cutting speed of 180 mm/min and feed/tooth of 0.062 mm/tooth were used for the machining cycle, and the material used is WAAM AISI 1040 steel. The axial depths of cut used for roughing, semi-finishing, and finishing were 2 mm, 1 mm, and 0.5 mm, respectively. In the finishing phase, two strategies were used: (1) chatter-free machining by selecting a spindle speed, and (2) minimising forced vibration by selecting spindle speeds far from the resonances of the system. Among these two strategies, no chatter was observed in the first strategy, but chatter occurred in the second. In addition to that, the first strategy resulted in better surface finish, but for both strategies, the errors were similar, as shown in Figure 7.8 [25].

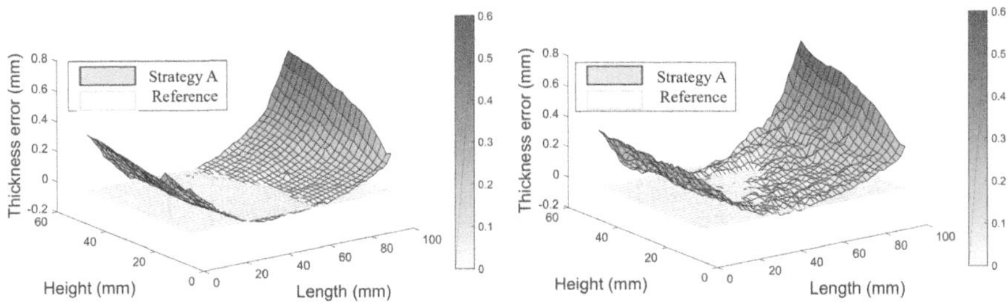

FIGURE 7.8 Thickness error for different strategies [25].

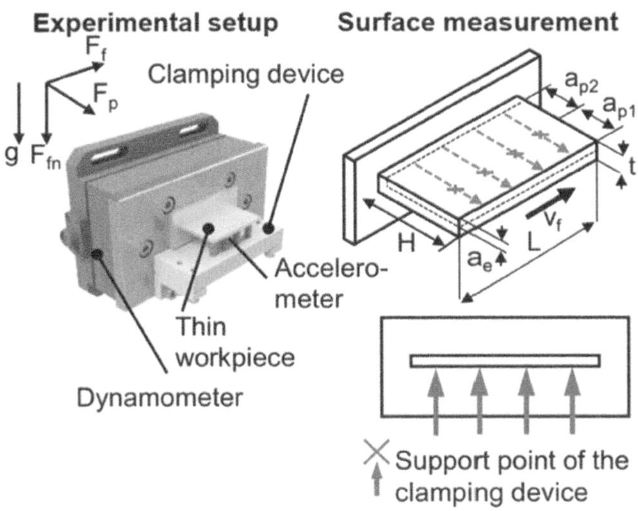

FIGURE 7.9 Machining strategy with support [26].

Another way to reduce the machining-induced deflection and inaccuracy in AM thin-walled structures is by using special support during milling, shown in Figure 7.9 [26]. Milling of thin-walled structures without any support results in deflection of the structure and an unmachined region at the end of the structure due to bending moments [26]. However, when milling at the middle of the structure, deflection and unmachined regions are reduced [26]. On the other hand, from Figure 7.10, when milling is carried out without support, a large portion of the unmachined layer was left on the edge of the thin-walled structure, and variation in cutting force and feed force was observed with changes in length [26]. From Figure 7.10, in the case of milling with support, uniform cutting and feed forces were observed, with a smaller portion of the unmachined layer left on the edge [26]. Furthermore, from Figure 7.11, it was observed that the amplitude of vibration is significantly reduced in the case of milling with support [26].

In some applications, thin-walled structures have internal microstructures. Machining such structures does not damage the microfeatures, but the machined chips occupy the space of the microfeatures. However, deviations in diameter were observed after machining thin-walled cylindrical parts with internal microfeatures.

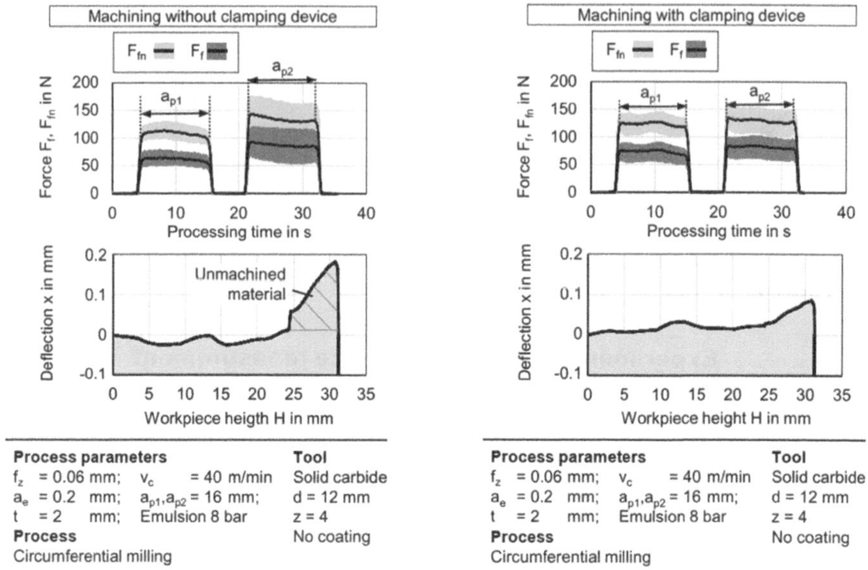

FIGURE 7.10 Machining forces and deflection with and without support [26].

The observed deviations are due to the residual stress in the printed parts and machining-induced compression to the outer walls [1]. The surface roughness was reduced after climb milling (R_a, 0.2–0.3 μm) compared to conventional milling (R_a, 0.3–0.7 μm). However, both milling conditions produce rough surfaces (R_a, 2–7.8 μm) in the corners of small radii. Additionally, in the case of conventional milling, the vibrations are lower; on the other hand, chip deposition in the cutting zone results in a poor finish [23]. Furthermore, the AM thin-walled part after machining (turning) shows a higher reduction in surface roughness (79%) and an improvement in average peak strength (6.5%) compared to the AB SLM-printed stainless steel [27].

In general, sandblasting is used to improve the surface quality of AM parts. Sandblasting of AM thin-walled parts shows a 43% improvement in surface finish and a negligible effect on the average peak strength [27]. Especially for finishing internal features, the abrasive flow finishing process is widely used. The abrasive flow finishing of thin-walled parts results in a 40% improvement in surface finish for large-sized channels, whereas for small-sized parts, a 38% improvement in surface finish was observed. In addition, a 6% change in diameter and attachment of abrasive particles was observed [28].

7.4.2.2 Surface Modification without Material Removal

The AB AM thin-walled structure has poor surface integrity, such as porosity, rough finish, residual stress, and anisotropy. Generally, surface post-treatments without material removal, such as peening and laser surface remelting, are carried out for the following purpose:

- To relieve the residual stress
- To increase the strength and hardness

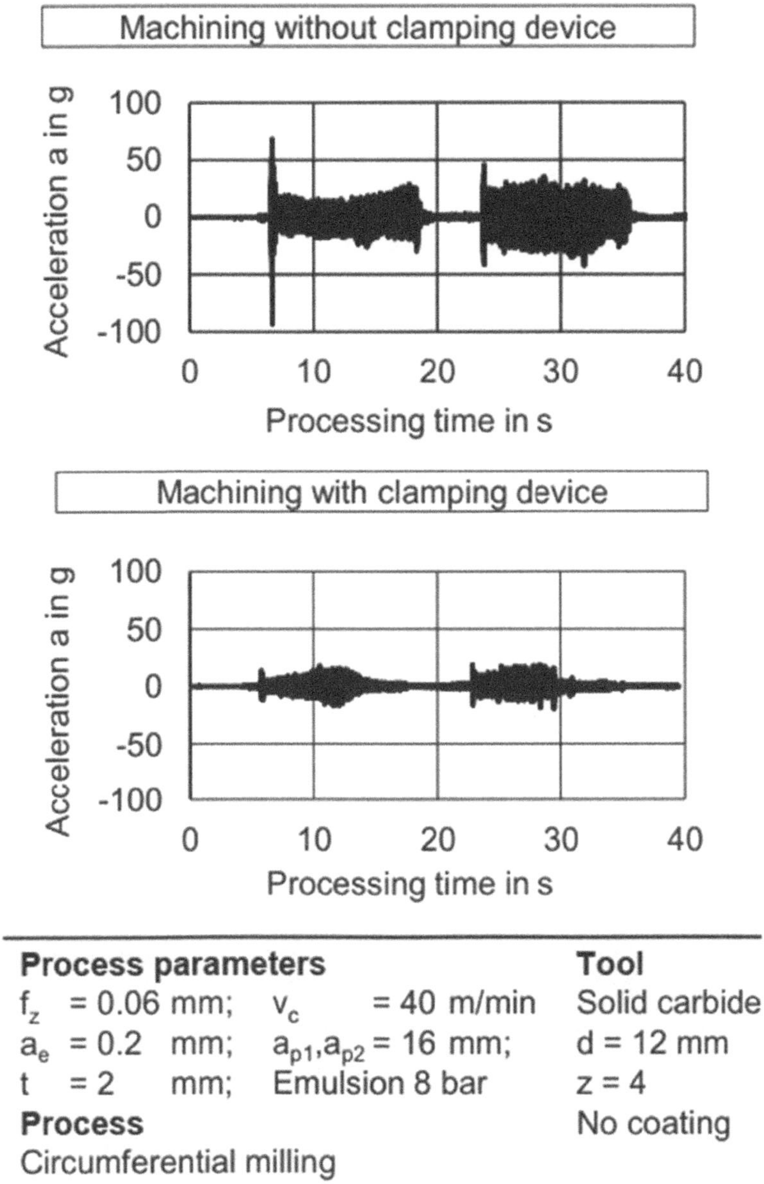

FIGURE 7.11 Amplitude of vibration during milling with and without support [26].

- To reduce the porosity
- To refine the microstructure

The LR process has the capability to modify surface integrity and can also be integrated into the printing machine itself. During the LR process, the peaks are transformed into valleys, resulting in a better surface finish. The surface finish obtained

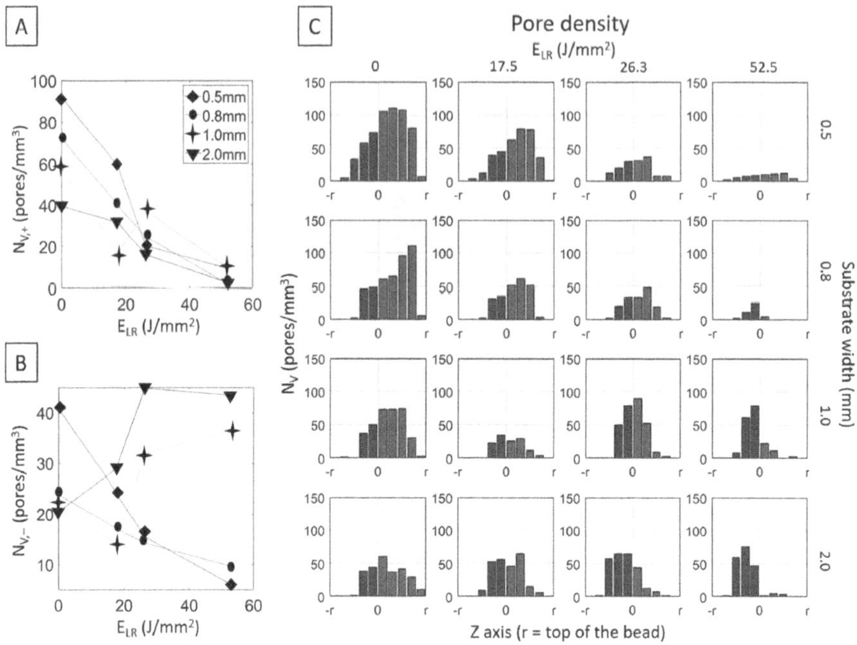

FIGURE 7.12 Scatter plots and histograms of pore density of single-track deposits. (A) Pore density in the top region of the deposited layer with respect to energy density of laser in four different substrate widths. (B) Changes in pore density in the bottom region of the deposited layer with respect to energy density of laser in four different substrate widths. (C) Histograms of pore density in the upper half (orange) and lower half (blue) of single-track deposits at different energy densities of laser and width [30].

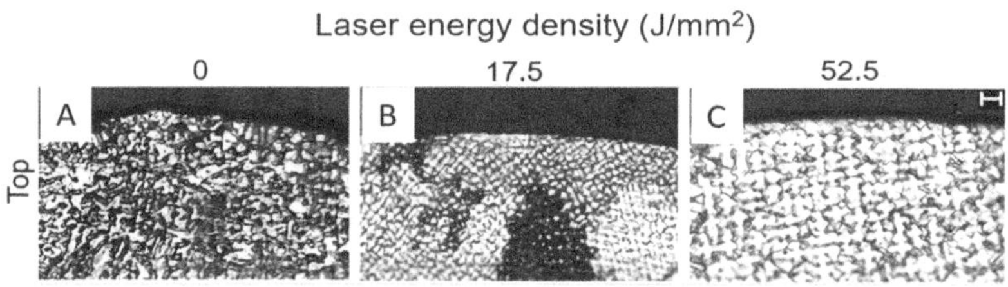

FIGURE 7.13 Change in microstructure with respect to energy density [30].

by the LR process depends on the energy density. When the energy density is high enough to remelt the surface of the AB AM thin-walled part, it results in a better surface finish. Overall, an 83% reduction in surface roughness and a 58% reduction in waviness were observed [29]. Additionally, LR reduces the pore density (up to 88%), as shown in Figure 7.12 [30], and significant changes in microstructure were observed from Figure 7.13 for varying energy density [30].

The peening process is generally used to improve the surface integrity of thin-walled parts. Different types of peening processes, such as ultrasonic peening,

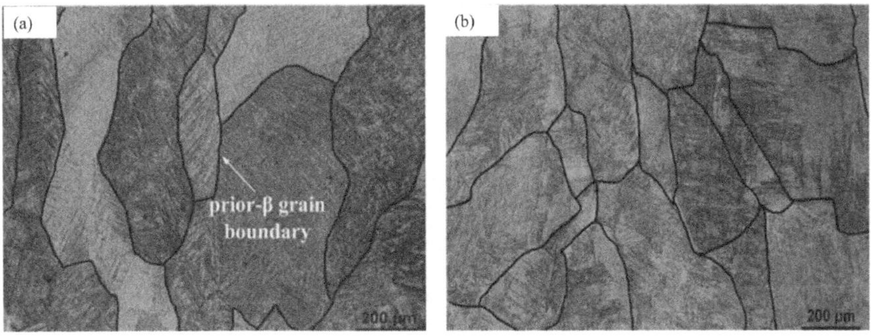

FIGURE 7.14 Microstructure (a) before UPT and (b) after UPT [31]. UPT, Ultrasonic peening treatment.

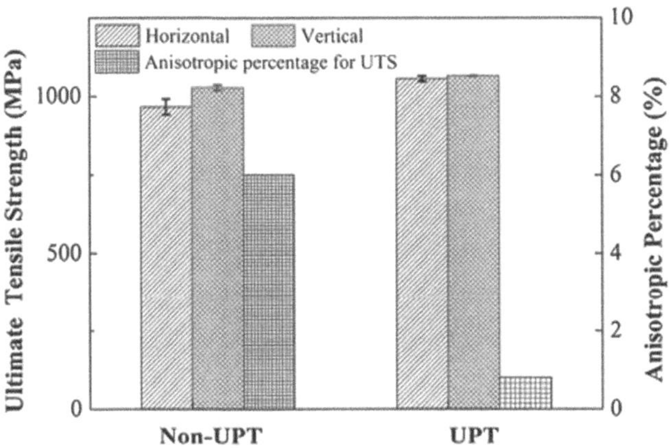

FIGURE 7.15 Ultimate tensile strength and anisotropy (a) before UPT and (b) after UPT [31]. UPT, Ultrasonic peening treatment.

LSP, and shot peening, are used for AM thin-walled parts. The ultrasonic peening treatment (UPT) reduces the grain size and causes grain refinement, as shown in Figure 7.14 [31]. Also, from Figure 7.15, it was observed that after UPT, the UTS in both directions increases, and the anisotropy is reduced [31]. On the other hand, LSP reduces residual stress by 200% [32] and porosity by 70%, as well as increases hardness by 49% compared to the thin-walled AB SLM AlSi10Mg part. Moreover, grain refinement occurs in the case of LSP, but minimal damage to surface roughness was observed. Owing to the grain refinement, from Figure 7.16, a 39.5% increase in tensile strength was observed after LSP treatment of AB WAAM thin-walled Al6Mg0.3Sc alloy [33]. The surface texture is also an equally important surface property, which generally gets affected during LSP treatment without coating due to ablation. However, LSP treatment with coating results in minimal damage to surface roughness and waviness, as shown in Figure 7.17 [34]. Apart from the above-mentioned peening processes, micro shot peening has less impact on strength [22].

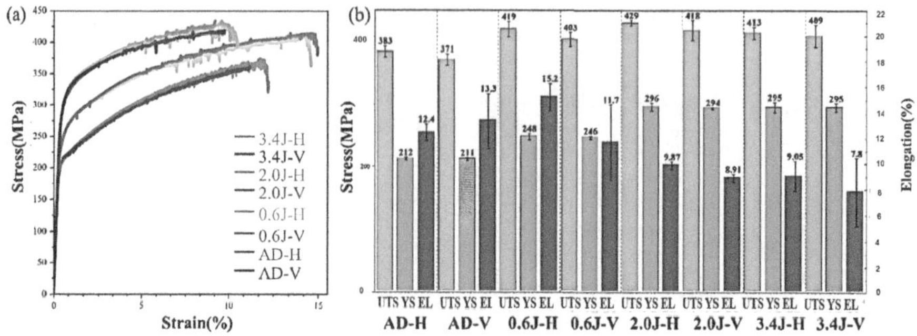

FIGURE 7.16 Mechanical behaviour before and after LSP (a) stress–strain curve (b) tensile strength and elongation with respect to energy [33] (The image published by Xuewei et al., 2024, is licensed under CC BY 4.0). LSP, Laser shock peening.

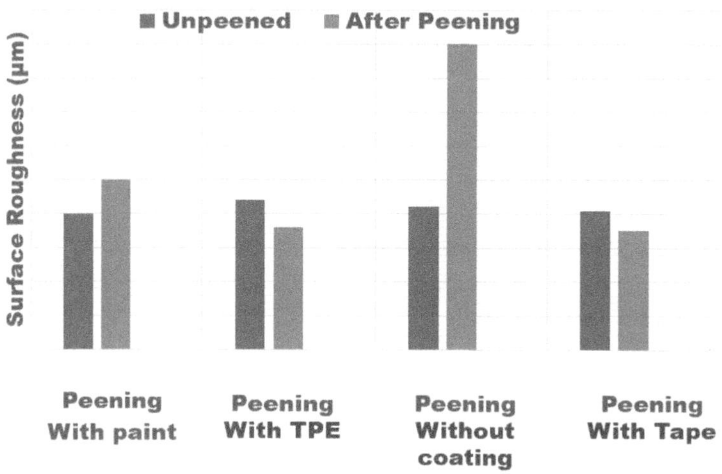

FIGURE 7.17 Effect of laser shock peening with and without coating on surface roughness.

7.4.2.3 Surface Modification by Adding Material

Applications like heat exchangers, turbine blades, bio-implants, landing gears, and so on require material addition or coating to provide the following functional performance:

- Wear resistance
- Corrosion resistance
- Biocompatibility
- Wettability

The above-highlighted functional performance can be obtained in the AB AM thin-walled structures by coating or material addition. In addition, damaged thin-walled turbine blades can be repaired by material addition processes such as electrodeposition [35]. Higher pulse energy and power inputs result in more defects and an increase in the width of the deposited layer. On the other hand, low pulse energy and power inputs result in voids and finer columnar subgrains, as well as higher microhardness [35]. Also, coating the AM part with tantalum via a sputtering process can improve corrosion resistance and biocompatibility [36].

7.5 APPLICATIONS OF ADDITIVE MANUFACTURED THIN-WALLED STRUCTURES

The AM thin-walled structures are used in marine, automobile, air vehicle, medical, electronics, and space applications. Some of the complex thin-walled structures made by additive manufacturing processes are listed below:

- Space applications – nozzle, thrust chamber, combustion liner, combustion chamber, structural jackets, fuel injector, space mirrors, solar panels, stator for liquid oxygen pump, and so on [4,6,37–39].
- Aircraft applications – seat brackets, actively cooled thrusters, combustion chambers, turbine blades, blisks, LEAP engine fuel injector nozzles, propellers, heat exchangers, thin-walled casings, and so on [4,5,39–41].
- Automobile applications – steering knuckle, thin-walled tubes, cooling vents, pumps, valves, power train, drive train, air duct, brake calliper, aluminium alloy water pump wheels for race cars, frame, intake manifold, integrated cooling channels, seat brackets, and so on [4,16,40,42].
- Energy applications – gas burner, burner head, blades in impellers, gas turbines, compressors, and so on [40].
- Medical applications – mandibular prosthesis scaffold, knee joint, femoral head, skull scaffolds, and so on [16].
- Electronics and marine – heat sinks, propellers, and so on [4,43–45].

In addition to the above applications, the DED process can be used for repairing damaged thin-walled structures, like turbine blades, impellers, propellers, and so on [16,35].

7.6 FUTURE TRENDS IN POST-PROCESSING OF ADDITIVELY MANUFACTURED THIN-WALLED STRUCTURES

Additive manufacturing processes can make intricate or complex features. However, to functionalise the complex thin-walled structure made by AM processes requires modifications or innovations in the post-processing stage. On the other hand, the process should be environmentally friendly in nature and energy efficient; also, the cost as well as the time for processing should be less. Thus, developing novel automatic post-processing methods is very important. The quality of the AM thin-walled structure can also be improved by modifying the design, proper support structure, and optimisation of process parameters. Besides, a proper understanding of process physics by using sensors and cameras can reduce the issues related to the AM thin-walled structures. Nowadays, sensor and camera data are used as input for developing AI models. The developed AI models predict the quality of the AM part. This will be beneficial for planning the subsequent operations. Another one of the recent interesting topics among the additive manufacturing research community is manufacturing in space. There is no considerable research on post-processing of thin-walled structures in microgravity and zero-gravity conditions.

The future trends in this domain are expected to focus on three key areas: (1) the development of advanced and automated post-processing methods specifically tailored for thin-walled AM structures, (2) the advancement of AI-driven models for real-time quality prediction and adaptive process control, and (3) post-processing studies on AB AM thin-walled parts in microgravity and zero gravity.

7.7 CONCLUSIONS

The microstructural post-treatments increase the strength of the AB AM thin-walled structures with excellent EA capability. In addition to that, the reduction in anisotropy and microstructural refinement was obtained. Likewise, the residual stress was relieved. On the other hand, the surface post-treatment, such as machining, produces a good surface finish, but to achieve considerable geometrical accuracy, a support is required. In contrast, the quality of the surface after remelting and peening depends on the energy density. The porosity and residual stress were reduced after surface remelting and peening treatment. Along with that, the strength also increased. However, especially in the case of LSP, minimal damage to surface roughness occurred. The electrodeposition can repair the damaged thin-walled structure by optimising the energy pulse and power input. Besides, the sputter coating of tantalum on AM parts can improve the corrosion resistance properties and biocompatibility.

Acknowledgements: The authors would like to acknowledge the Department of Mechanical Engineering, SRM Institute of Science and Technology, Vadapalani, Chennai-26, India, for the support.

REFERENCES

1. Mishra, Sarvesh Kumar, Gaizka Gómez-Escudero, Haizea González-Barrio, Amaia Calleja-Ochoa, Silvia Martinez, Michale Barton, and Luis Norberto Lopez De Lacalle. "Machining-induced characteristics of microstructure-supported LPBF-IN718 curved thin walls." *Procedia CIRP* 108 (2022): 176–181.
2. Niu, Xu, Ruixian Qin, Yunzhuo Lu, and Bingzhi Chen. "Energy absorption behaviors of laser additive manufactured aluminium alloy thin-walled tube tailored by heat treatment." *Mater Trans* 62 (2021): 278–283.
3. Sinha, Ayush, Biswajit Swain, Asit Behera, Priyabrata Mallick, Saswat Kumar Samal, H. M. Vishwanatha, and Ajit Behera. "A review on the processing of aero-turbine blade using 3D print techniques." *Journal of Manufacturing and Materials Processing* 6 (2022): 1–34.
4. Viveganathan, Priyanka, Shubo Gao, Weiming Ji, Haiyang Fan, Changjun Han, and Kun Zhou. "Recent progress on 3D printing of lightweight metal thin-walled structures." *Advanced Materials* 36(2024): 2402130–2402170.

5. Careri, Francesco, Raja H.U. Khan, Catherine Todd, and Moataz M. Attallah. "Additive manufacturing of heat exchangers in aerospace applications: A review." *Applied Thermal Engineering* 235 (2023): 121387.
6. Kerstens, Fabio, Angelo Cervone, and Paul Gradl. "End to end process evaluation for additively manufactured liquid rocket engine thrust chambers." *Acta Astronaut* 182 (2021): 454–465.
7. Asadi, Reza, Antoine Queguineur, Olli Wiikinkoski, Hossein Mokhtarian, Tommi Aihkisalo, Alejandro Revuelta, and Iñigo Flores Ituarte. "Process monitoring by deep neural networks in directed energy deposition: CNN-based detection, segmentation, and statistical analysis of melt pools." *Robotics and Computer-Integrated Manufacturing* 87 (2024): 102710–102724.
8. Abhilash, Puthanveettil Madathil, and Afzaal Ahmed. "Convolutional neural network-based classification for improving the surface quality of metal additive manufactured components." *International Journal of Advanced Manufacturing Technology* 126 (2023): 3873–3885.
9. Shunmugasamy, Vasanth C., Eisha Khalid, and Bilal Mansoor. "Friction stir extrusion of ultra-thin wall biodegradable magnesium alloy tubes—Microstructure and corrosion response." *Materials Today Communications* 26 (2021): 102129–102141.
10. Lazoglu, Ismail, and Ali Mamedov. "Deformation of thin parts in micromilling." *CIRP Annals of Manufacturing Technology* 65 (2016): 117–120.
11. Ding, Donghong, Runzhuo Zhao, Qinghua Lu, Zengxi Pan, Huijun Li, Kai Wang, and Kuanfang He. "A shape control strategy for wire arc additive manufacturing of thin-walled aluminium structures with sharp corners." *Journal of Manufacturing Processes* 64 (2021) :253–264.
12. Lv Wei, Mei Zhan, Pengfei Gao, Ming Li, Yudong Lei, and Fei Ma. "Improvement of rib-grid structure of thin-walled tube with helical grid-stiffened ribs based on the multi-mode filling behaviors in flow forming." *Journal of Materials Processing Technology* 296 (2021): 117167–117179.
13. Zhang, Lijun, Lukuan Li, Ning Yang, Su Liu, Hong Chi, Jiayi Xu, Shuai Zhang, Fazhe Sun, Zhong Jian, He Qingqiang, and An Yulei. "Active shovel spinning process: Plastic deformation behavior, microstructure, and properties." *Thin-Walled Structures* 206 (2025): 112714–112728.
14. Körber, Selina, Rainer Völkl, and Uwe Glatzel. "3D printed polymer positive models for the investment casting of extremely thin-walled single crystals." *Journal of Materials Processing Technology* 293 (2021): 117095–117103.
15. Lee, Jian Yuan, Arun Prasanth Nagalingam, and Swee Hock Yeo. "A review on the state-of-the-art of surface finishing processes and related ISO/ASTM standards for metal additive manufactured components." *Virtual Physical Prototyping* 16 (2021): 68–96.
16. Liu, Shunyu, and Yung C. Shin, "Additive manufacturing of Ti6Al4V alloy: A review." *Materials & Design* 164 (2019): 107552–107575.
17. Majeed, Arfan, Muhammad Muzamil, Jingxiang Lv, Bufan Liu, and Fiaz Ahmad. "Heat treatment influences densification and porosity of AlSi10Mg alloy thin-walled parts manufactured by selective laser melting technique." *Journal of the Brazilian Society of Mechanical Sciences and Engineering* 267 (2019):1–13.
18. Grigoriev, Sergey N., Nikita Nikitin, Oleg Yanushevich, Natella Kriheli, Olga Kramar, Roman Khmyrov, Idarmach Idarmachev, and Pavel Peretyagin. "Experimental and statistical analysis of the effect of heat treatment on surface roughness and mechanical properties of thin-walled samples obtained by selective laser melting from the Material AlSi10Mg" *Materials* 16 (2023): 7326–7349.

19. Zhang, Jian, Jikang Fan, Linye Duan, Wei Chen, Dongqing Yang, and Kehong Wang. "Effect of heat treatment on the microstructure and mechanical properties of 18Ni 300 maraging steel manufactured by wire + arc additive manufacturing." *Virtual and Physical Prototyping* 19 (2024): 1–19.
20. Banerjee, Arunima, Mo Rigen He, William D. Musinski, Paul A. Shade, Marie E. Cox, Edwin J. Schwalbach, and Kevin J. Hemker. "Effect of stress-relief heat treatments on the microstructure and mechanical response of additively manufactured IN625 thin-walled elements." *Materials Science and Engineering: A* 846 (2022): 143288–143315.
21. Colón, Brandon, Mehrdad Pourjam, Gabriel Demeneghi, Kavan Hazeli, Omar Mireles, and Francisco Medina. "On the mechanical behavior of LP-DED C103 thin-wall structures." *Metals (Basel)* 14 (2024): 958–979.
22. Spignoli, Nicola, and Giangiacomo Minak. "Influence on fatigue strength of post-process treatments on thin-walled AlSi10Mg structures made by additive manufacturing." *Metals (Basel)* 13 (2023): 126–138.
23. Isaev, Alexander, Vladimir Grechishnikov, Petr Pivkin, Kozochkin Mihail, Yuriy Ilyuhin, and Andrey Vorotnikov. "Machining of thin-walled parts produced by additive manufacturing technologies." *Procedia CIRP* 41 (2016): 1023–1026.
24. Moriya, Shinichi, Takuma Inoue, Masahiro Sasaki, Takayuki Nakamoto, Takahiro Kimura, Naoyuki Nomura, Keiko Kikuchi, Akira Kawasaki, Toshiki Kato, and Ideo Masuda. "Feasibility study on additive manufacturing of liquid rocket combustion chamber." *Transactions of the Japan Society for Aeronautical and Space Sciences, Aerospace Technology Japan* 16 (2018): 261–266.
25. Grossi, Niccolò, Antonio Scippa, Giuseppe Venturini, and Gianni Campatelli. "Process parameters optimization of thin-wall machining for wire arc additive manufactured parts." *Applied Sciences (Switzerland)* 10 (2020): 7575–7588.
26. Hintze, Wolfgang, Robert von Wenserski, Sebastian Junghans, and Carsten Möller. "Finish machining of Ti6Al4V SLM components under consideration of thin walls and support structure removal." *Procedia Manufacturing* 48 (2020): 485–491.
27. Mehta, Bharat, Eduard Hryha, Lars Nyborg, Frederic Tholence, and Erik Johansson. "Effect of surface sandblasting and turning on compressive strength of thin 316L stainless steel shells produced by laser powder bed fusion." *Metals (Basel)* 11 (2021): 1070–1084.
28. Jalui, Sagar, T. J. Spurgeon, Harold Jacobs, Arnab Chatterjee, Timothy Stecko, and Guha Manogharan. "Abrasive flow machining of additively manufactured titanium: Thin walls and internal channels." *Proceedings of the 32nd Annual International Solid Freeform Fabrication Symposium*, pp. 1646–1660, 2021.
29. Bruzzo, Francesco, Guendalina Catalano, Ali Gökhan Demir, and Barbara Previtali. "In-process laser re-melting of thin-walled parts to improve surface quality after laser metal deposition." *Key Engineering Materials* 813 (2019): 191–196.
30. Kim, Myong Joon, and Christopher Saldana. "Post-processing of additively manufactured IN625 thin-walled structures using laser remelting in directed energy deposition." *Journal of Manufacturing Processes* 88 (2023): 59–70.
31. Gou, Jian, Zhijiang Wang, Shengsun Hu, Junqi Shen, Yinbao Tian, Guancheng Zhao, and Youquan Chen. "Effects of ultrasonic peening treatment in three directions on grain refinement and anisotropy of cold metal transfer additive manufactured Ti-6Al-4V thin wall structure." *Journal of Manufacturing Processes* 54 (2020): 148–157.
32. Stránský, Ondřej, Libor Beránek, Sunil Pathak, Jan Kaufman, Marek Böhm, Jaromír Kopeček, František Holešovský, Šimon Petrášek, Lucie Hlavůňková, and Zbyněk Soukup. "Porosity and microstructure of L-PBF printed AlSi10Mg thin tubes in laser shock peening." *Journal of Materials Research and Technology* 27 (2023): 1683–1695.

33. Fang, Xuewei, Kai Li, Minghua Ma, Chang Liu, Yao Zhang, Jian Zhou, Zhanxin Li, Jiahao Shang, Ke Huang, and Bingheng Lu. "Improved properties of wire arc directed energy deposited thin-walled Al-6Mg-0.3Sc component via laser shock peening." *Virtual and Physical Prototyping* 19 (2024): 1–18.
34. Stránský, Ondřej, Libor Beránek, Sunil Pathak, Jan Šmaus, Jaromír Kopeček, Jan Kaufman, Marek Böhm, Jan Brajer, Tomáš Mocek, and František Holešovský. "Effects of sacrificial coating material in laser shock peening of L-PBF printed AlSi10Mg." *Virtual and Physical Prototyping* 19 (2024): 1–15.
35. Yan, Jihui, Pablo D. Enrique, Kevin Chan, Nigel Scotchmer, Peng Peng, and Norman Y. Zhou. "Repair of additive manufactured thin-wall geometry using electrospark deposition." *Journal of Remanufacturing* 13 (2023): 333–353.
36. Sui, Zehui, Jian Wang, Cong Wu, Jingzhe Niu, Jianfeng Zhu, and Lian Zhou. "Research on the surface characterization, corrosion and bioactivity of nano-featured tantalum coating on selective electron beam melted Ti6Al4V alloy." *Journal of Alloys and Compounds* 946 (2023): 169351–169363.
37. Dordlofva, Christo, Angelica Lindwall, and Peter Törlind. "Opportunities and challenges for additive manufacturing in space applications." In the *Proceedings of NordDesign 2016*, Volume 1, Trondheim, Norway, 10th–12th August 2016 pp. 401–410, 2016.
38. Blachowicz, Tomasz, Guido Ehrmann, and Andrea Ehrmann. "Metal additive manufacturing for satellites and rockets." *Applied Sciences* 11 (2021): 12036–12051.
39. Gradl, Paul, Darren C. Tinker, Alison Park, Omar R. Mireles, Marissa Garcia, Ryan Wilkerson, and Christopher Mckinney. "Robust metal additive manufacturing process selection and development for aerospace components." *Journal of Materials Engineering and Performance* 31 (2022): 6013–6044.
40. Tepylo, Nick, Xiao Huang, and Prakash C. Patnaik. "Laser-based additive manufacturing technologies for aerospace applications." *Advanced Engineering Materials* 21(2019): 1900617–1900652.
41. Arya, Pradyumn Kumar, Neelesh Kumar Jain, Dan Sathiaraj, and Vivek Patel. "Development of high strength and lightweight Ti6Al4V5Cr alloy: Microstructure and mechanical characteristics." *Journal of Materials Research and Technology* 28 (2024): 3526–3540.
42. Yang, Jian, Bo Li, Jian Liu, Zhantong Tu, and Xin Wu. "Application of additive manufacturing in the automobile industry: A mini review." *Processes* 12 (2024): 1101–1127.
43. Ziółkowski, Marcin, and Tomasz Dyl. "Possible applications of additive manufacturing technologies in shipbuilding: A review." *Machines* 8 (2020): 1–34.
44. Biao, Zhao, Ding Wenfeng, Shan Zhongde, Wang Jun, Yao Changfeng, Zhao Zhengcai, Liu Jia, Xiao Shihong, Ding Yue, Tang Xiaowei, Wang Xingchao, Wang Yufeng, and Wan Xin. "Collaborative manufacturing technologies of structure shape and surface integrity for complex thin-walled components of aero-engine: Status, challenge and tendency." *Chinese Journal of Aeronautics* 36 (2023): 1–24.
45. Pradyumn Kumar Arya, Pankaj Kumar, Balbir Singh Negi, Neelesh Kumar Jain, and Dan Sathiaraj. "Corrosion behaviour of Ti6Al4VxCryNi alloys fabricated by µ-plasma arc powder additive manufacturing process." *Journal of Alloys and Compounds* 984 (2024): 173980–173992.

8 Post-Processing Techniques for Metal Additively Manufactured Thin-Walled Structures

Shubhavardhan Ramadurga Narasimharaju, Andrea El Hassanin, Fabio Scherillo, Antonino Squillace, Antonio Viscusi, Abhijit Cholkar, Ramesh Raghavendra, and Dermot Brabazon

8.1 INTRODUCTION

Since its inception, dating back to 1986, additive manufacturing (AM) has been developed and patented from a stereolithography machine [1]. As per the International Organisation for Standardisation (ISO) and American Society for Testing and Materials (ASTM), ISO/ASTM 52900; AM is the addition of minuscule layers of feedstock material consolidation to build three-dimensional (3D) physical objects [1], which is contrary to the conventional subtractive manufacturing processes [2]. AM, commonly referred to as 3D printing, is gaining more popularity amongst academia, research, and the industrial fraternity for functional parts production while offering greater design freedom [2–4], which enables manufacturing geometrically complex parts directly from computer-aided design files that traditional manufacturing processes cannot produce [3,4]. In addition, AM helps to reduce material consumption, chemical waste, process steps, and human resources [5]. Driven by these unique characteristics, AM has in recent years made an outstanding contribution to the design and fabrication of intricate functional components used in biomedical, aerospace, automotive, electronics, electrical, shipbuilding, consumer goods, and defence industrial applications [6–8].

Despite AM being capable of building the most complicated dimensional parts, which in most cases are considered impossible in traditional processes, the surface quality, also known as surface finish or surface texture, is one critical limitation or a key challenge that is extremely worth overcoming for the

 DOI: 10.1201/9781003571209-8

rapid progress of the metal AM process [9]. Additionally, the post-processing of complex AM geometries such as lattice structures, internal cooling channels, freeform surfaces, etc., poses extreme challenges compared to other components [2,9].

In general, most of the properties of the materials are indeed considered the standard surface properties of mechanical products: for example, fatigue, cracking, friction and wear, oxidation, abrasion, adhesion and bonding, erosion, etc., are all influenced by the surface properties, as surface quality plays a vital role in determining the final part quality and functional performance of many engineering applications, including metal AM parts [10–14]. The surface quality contributes up to 10% of the part failure rate in conventionally fabricated parts, but this is estimated to be even higher in the case of metal AM components [15]. Some of the major contributors accountable for poor surface quality or rougher surface finish of AM parts are: firstly, the AM process itself, which induces higher surface roughness as it undergoes a very complex thermophysical mechanism and kinetic phenomena during the layer-by-layer building process [16]. Secondly, the right selection amongst the diverse AM process parameters plays its part in imparting a unique anisotropic fingerprint on surface features, generating poor surface quality [17]. Finally, miscellaneous factors such as lower process resolution of building part/artefact and layering effects, supporting structures, feedstock materials quality, packing density, morphology, and rheology contribute towards the substandard surface quality [18]. Therefore, the final part quality of AM parts is perceived to be unpredictable and uncontrollable. The individual or the combination of these circumstances yields strong surface topographical characteristics inherent to the AM process, including the metal AM-processed components, usually dominated by the number of metallurgical defects and surface asperities [2,8,9] (see Figure 8.1). The most commonly occurring metallurgical AM defects are porosities, lack of fusion holes, thermal cracks, oxidation, balling or humping, ripple effect, denudation, loss of alloying elements, and the surface asperities, such as spatters, unmelted/partially melted powder particles, re-entrant features, surface and sub-surface porosities, and staircase effect [19]. The anisotropic microstructure and elaborate surface topography induced by the AM process are not trivial and present a new challenge for conventional metrology. Also, these inevitable defects and asperities induce substandard surface quality and compromise the dimensional accuracy of intricate AM components, which are constantly failing to meet the specifications of functional industrial standards [9]. It is paramount to characterise, correlate, and quantify these defects and asperities, as they eventually determine the long-term functional performance of as-built AM components [8]. For instance, the emergence of these defects and asperities adversely affects the stress distribution under mechanical loading conditions, resulting in premature failure of AM components [20]. However, the overall quality of intricate AM structures like thin-walled structures and/or lattice structures can be improved by using the optimised design or by employing various advanced post-processing treatment methods, but ascertaining the suitable post-processing technique consumes more time, money, and energy [2,8,9].

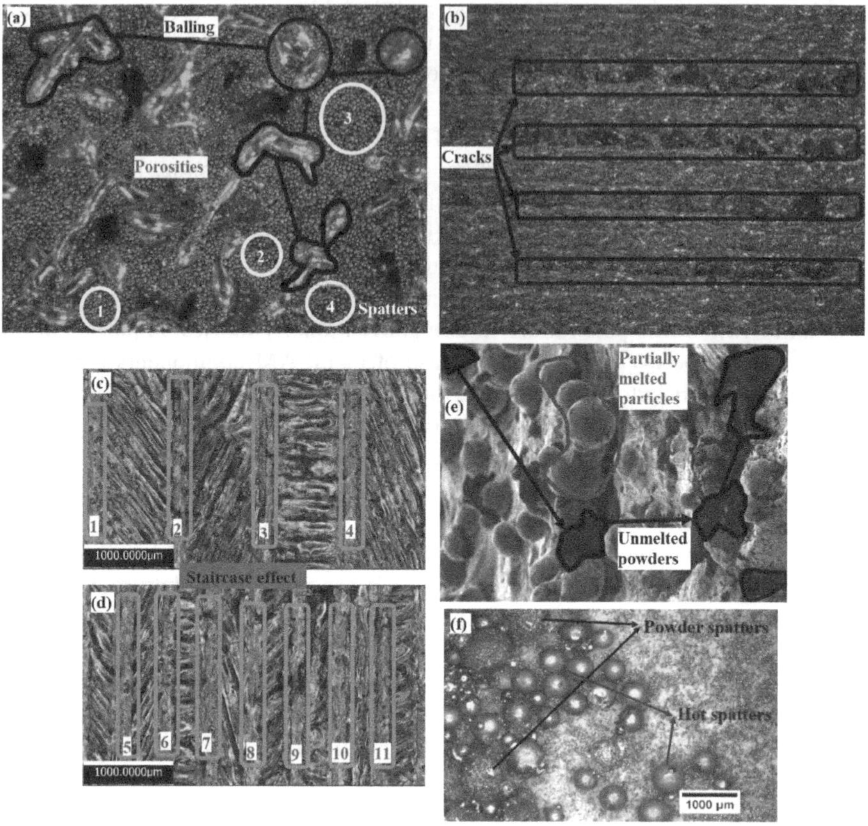

FIGURE 8.1 Schematic illustration of metallurgical defects and surface asperities of intricate AM components [2].

8.2 IMPORTANCE OF ADVANCED POST-PROCESSING TECHNIQUES FOR AM PARTS: THIN-WALLED STRUCTURES

The advanced post-processing treatment methods aid in improving the overall performance of the AM final part quality and promote their applications to fulfil specified industrial standards. The as-built AM components, including thin-walled structures, must undergo one or more forms of post-processing treatments to fulfil these functional specifications. Hot-isostatic pressing (HIP), shot peening (SP), chemical polishing, electrochemical polishing, continuous wave (CW) laser hardening, laser polishing, vapour polishing, mechanical polishing, and other hybrid methods are the popular post-processing techniques applied for complex AM parts [19–22]. The industrially recognised HIPing helps to increase the part density while reducing the number of porosities and refining the microstructure. SP, mechanical polishing, and hybrid methods are used to significantly improve the surface quality of as-built AM parts. Overall, the post-processing treatments help to improve the holistic (internal/external) final part quality [21,22].

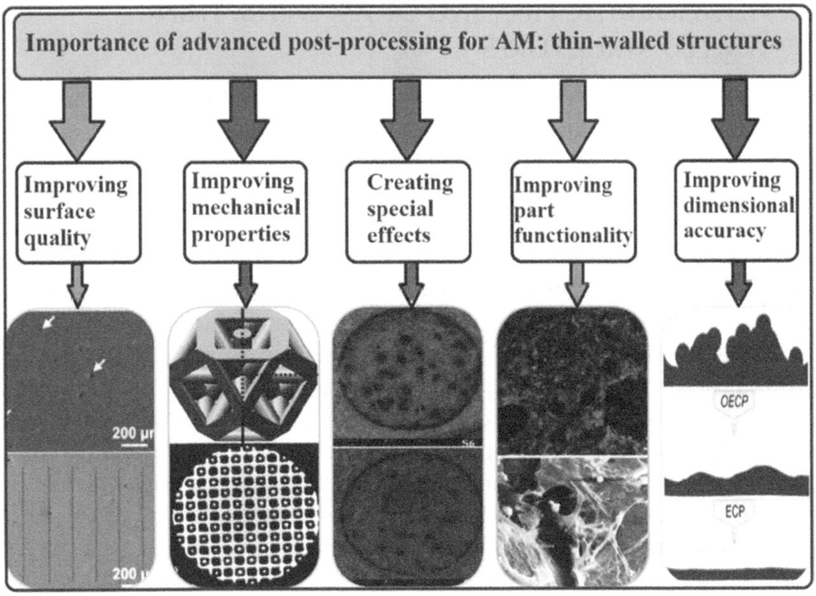

FIGURE 8.2 Schematic illustration of the importance of advanced post-processing AM components [21].

The chemical and electrochemical post-processing methods are usually effective in removing adhered powders from internal cavities and simultaneously minimising the surface roughness, resulting in increased dimensional accuracy [21]. The chemical surface finishing and similar treatments are applied to enhance the corrosion resistance and surface hardness of complex AM parts like thin-walled structures. The mechanical post-processing techniques are less popular for the complex AM thin-walled structures and lattice parts [22]. The primary importance of advanced post-processing techniques for AM components, including thin-walled structures, is to improve surface quality and to enhance the mechanical properties of as-built parts. Additionally, these advanced methods can also be employed to create special effects, enhance part aesthetics, functionality, and finally, be useful to improve dimensional accuracy (see Figure 8.2) [21,22]. The heat treatment post-processing methods, such as quenching or ageing, help to improve, tailor, or achieve the desired mechanical properties of metal AM components by promoting a uniform crystal structure, thereby improving their ultimate tensile strength and hardness [19]. Heat treatments and controlled cooling help to improve the AM part shrinkage rate, thereby helping to maintain the dimensional stability [21]. Creating special effects herein refers to increasing corrosion resistance of titanium alloy by anodising the oxide layer, or changing colour, etc., for as-built AM parts [21]. Disinfection and sterilisation are some of the popular examples of post-processing methods used to provide additional functionality to very intricate AM parts [21,22].

8.3 HOT-ISOSTATIC PRESSING OF AM PARTS: THIN-WALLED STRUCTURES

HIP post-processing technique was developed in the mid-1950s to reduce porosities and further densify the consolidated powders of sintered metals, ceramics, or composite materials parts within high-temperature and isostatic high-pressure environments [23]. The HIP process involves the placement of the as-built AM part in a sealed container with inert gas. This sealed container is then positioned within a high-temperature (up to 2000°C) and high-pressure (up to 2700°C) vessel (see Figure 8.3). The processing temperature of HIP is usually below the melting point and higher than that of the solidus temperature line, as per ASTM F3301 [23]. During the HIP process, the as-built AM samples are subjected to mechanical pressing under isostatic pressure and elevated temperatures [23]. The mechanical pressing under isostatic pressure consolidates the metallic, ceramic, and composite material powders, resulting in enhanced AM part density, simultaneously eliminating the metallurgical defects of intricate AM parts' internal porosities, lack of fusion holes, microcracks, etc. [19,21]. HIP also reduces notch sensitivity of defects, lowers anisotropy through recrystallisation, relieves thermal stresses, and minimises

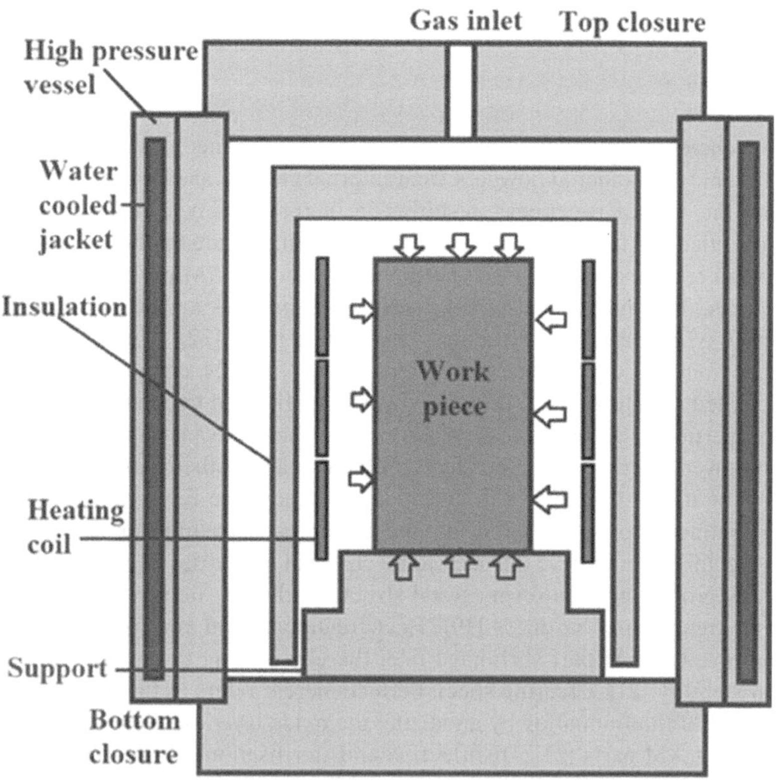

FIGURE 8.3 Schematic illustration of hot-isostatic pressing equipment [42].

chemical segregation while promoting the formation of equiaxed grains, leading to homogeneity of microstructure [23]. As a result, HIP-processed AM samples show higher strength, ductility, toughness, and corrosion resistance. The highly dense 3D AM parts with minimal defects and optimised mechanical properties have been widely used in reliable functional components of aerospace, biomedical, electrical, electronics, transport, marine, shipbuilding, and consumer goods industrial sector applications [22,23]. Some of the examples of very complex AM functional components already in service are rocket engine blades, thin-walled cylinders, turbine discs, rocket engine conforming cooling channels, guide vanes, pipeline connectors, motherboard circuits, dental, spinal, and similar ortho implants, surgical tools, marine shoes, etc. [19–23].

Currently, HIP is employed to post-process metal AM components (Ti alloy, stainless steel (SS), tool steel, Mg alloys, WC-Co alloys, CoCr, Al alloys, super alloys, Cu, and Inconel alloys) [24–42]. The isostatic pressing at elevated temperatures with sufficient time, creep and diffusion mechanisms result in pore elimination of residual gas entrapped in the original powder particles without transporting into the surface, displaying greater mechanical bonding with an overall increase in tensile properties (elongation ~9%) of the complex SS AM parts [24]. The reduction of pores, solid solution of the secondary phase, and part densification occur simultaneously, giving rise to a uniform microstructure; this combined effect promotes elongation and plasticity and overall improvement in the mechanical properties of stainless-steel AM components [24,25]. Owing to a uniform microstructure with minimal porosity, dislocation density, and continuous grain growth, the tensile ductility collectively increased by 10–20% [24,25]. Annealing twins are also evident in the microstructure of HIPed AM parts due to sufficient holding time at higher temperatures during operation [23,29,34]. HIP imparted the α grain coarsening, and the β grain formation led to improved toughness of the AM Ti alloy [26–28]. The homogeneous microstructure resisted higher fatigue loads for a prolonged period before fatigue crack initiation resulted in excellent fatigue strength of 450–500 MPa [28]. The improved fatigue strength (by 106 cycles) and endurance ratio (0.3–0.55) of HIPed Ti lattice structures are evident, which is equivalent to the wrought Ti materials [30]. The HIP transformed the uneven columnar grain microstructure to uniform recrystallised grains precipitated by carbides, which led to improved microhardness, Young's modulus, and the texture strength of AM-processed high-temperature Inconel 718 alloy [37]. HIP significantly reduced the micro-holes in as-built AM CrCo alloy, resulting in ~56% improvement in strain energy density. Additionally, HIP activated deformation twinning under low strain after HIP, enabling stable work hardening behaviour [29,34]. In summary, HIP is employed to eliminate the metallurgical defects, homogenise the microstructure, and increase the part density, thereby improving the overall mechanical properties of complicated AM parts, including thin-walled structures.

8.4 SHOT PEENING OF AM PARTS; THIN-WALLED STRUCTURES

SP post-processing treatment involves spraying high-speed small, spherical beads or hard particles (shots; metallic, glass, or ceramic) onto the sample's surface to generate

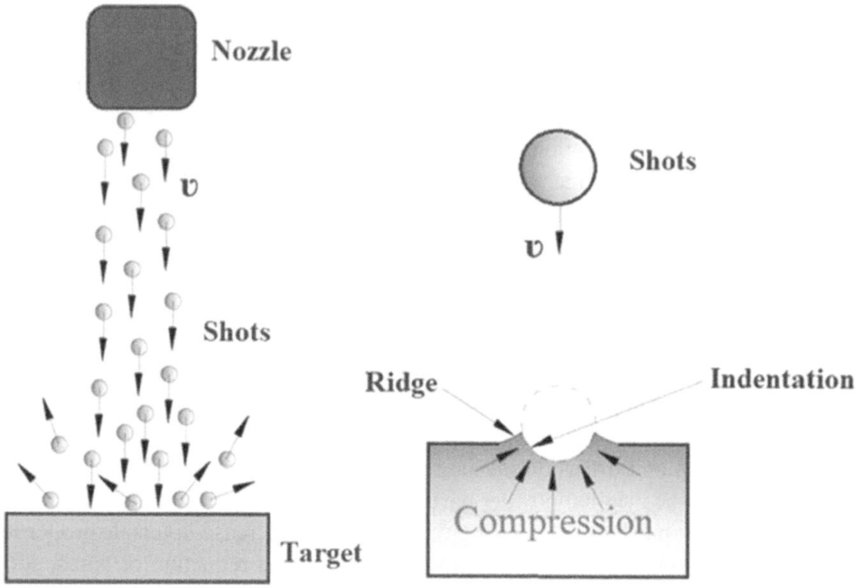

FIGURE 8.4 Schematic illustration of the shot peening process [44].

multiple impacts. The impact of these hard particles or shots with sufficient velocities ensures plastic deformation with minimal heat generation [43] (see Figure 8.4) [44]. The SP creates both transverse and nominal forces that cover multiple impacts for 100% coverage due to the spherical nature of the hard particle impacts, which results in increased hardness [43]. SP can be carried out with steel balls (S110, utilising an Almen intensity of I = 5`'A) or ceramic balls (Z210, utilising 12.2`N) to post-process AM metal (Ti, SS, Al, Inconel 718 alloy) components. SP significantly improves micro-nano hardness, increases corrosion resistance, minimises surface defects, and reduces surface roughness between 50 and 86% of AM components [23,45].

As a result of the hard particles' bombardment on the AM surfaces, a continuous smooth surface with a finer surface finish and minimal surface imperfections creates a compressive residual layer. This compressive residual layer effectively resists higher fatigue loads before crack initiation, resulting in achieving superior fatigue limits of SP AM Ti components [46]. SP generated a compressive stress layer of 400 μm for the AM-processed Inconel 718 component surface [43]. The surface roughness of shot-peened AM AlSiMg samples showed improvement in surface roughness by 86.6% (as-built; 30 μm, SPed; 4.6 μm). This significant improvement in surface finish greatly increased the fatigue limit by 120% (as built; 50 MPa, SPed; 110 MPa at 10^7 cycles) [47]. The SP treatment also helps to minimise the surface porosity defects by 0.1–0.3% on the complex AM surfaces with a surface area of up to 500 μm deep, which promotes increased fatigue strength (33.3%) at 10^7 cycles from 60 to 80 MPa [48]. All in all, SP post-process treatment is adopted to improve the surface roughness, minimise the surface defects, increase hardness, and create a compressive residual stress layer beneficial to significantly improve the mechanical properties, especially fatigue properties, of complex AM parts, including thin-walled structures.

The HIP and SP processes' working principle is not extensively discussed as it is widely available in the existing literature. However, the advanced post-processing techniques such as laser surface hardening (LSH), fluidised bed finishing (FBF), and chemical machining are discussed in detail.

8.5 LASER SURFACE HARDENING OF AM PARTS: THIN-WALLED STRUCTURES

LSH is an advanced laser-based material surface processing technique in which the surface of a workpiece is rapidly heated for a brief duration, metallic parts above their transformation temperature, using high-power laser radiation, and then followed by self-quenching due to the surrounding cooler material [49,50]. This localised energy input induces physical and/or chemical changes on the material's surface and enhances the surface properties of metals such as hardness, wear resistance, and other mechanical properties [51,52]. In this process, the rapid heating and cooling create a hardened surface layer while preserving the metal's bulk properties, thereby improving surface performance without compromising overall structural integrity. Therefore, amongst various surface enhancement techniques, LSH has emerged as a promising method due to its localised treatment, precision, and minimal distortion [53].

8.5.1 Working Principle of Laser Surface Hardening

A schematic flowchart of the mechanism of LSH is shown in Figure 8.5. In LSH, fluxes generated by lasers are most suitable to heat the surface layer to authentication levels without affecting the bulk temperature of the sample. The core principle of LSH relies on rapid heating and self-quenching of the material surface [54]:

1. A focused laser beam is directed onto the metal surface, where its energy is absorbed and converted into heat.
2. The surface temperature rises rapidly (10^3–10^6 °C/s), typically above the austenitisation temperature (e.g. 727–950 °C for steels), without melting the material.
3. Once the laser beam moves away, the underlying cooler bulk material acts as a heat sink, rapidly cooling the heated surface at rates sufficient to suppress diffusion-controlled transformations.
4. The surface undergoes a solid-state transformation (e.g. formation of martensite in steels), leading to a significant increase in hardness and wear resistance.

Figure 8.6 illustrates the temperature profile across the depth of the material and the laser-material interaction zone [55].

During LSH, a thermal gradient is established along the depth of the material, inducing different microstructural zones:

Laser-hardened zone: Located at the surface, transformed into a fine martensitic or bainitic structure.

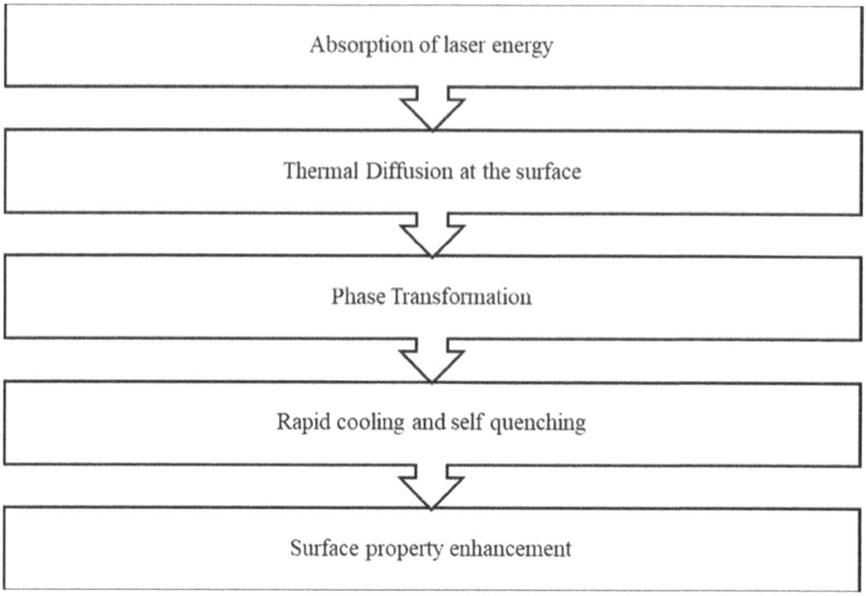

FIGURE 8.5 Schematic flowchart of the mechanism of laser surface hardening.

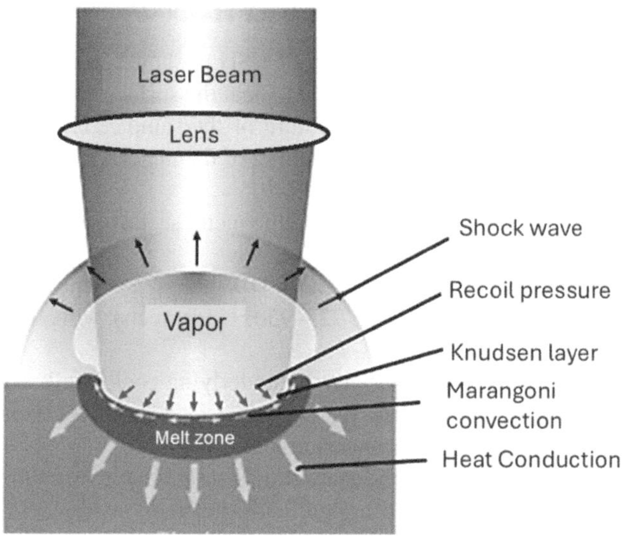

FIGURE 8.6 Schematic of laser surface hardening [55].

Heat-affected zone (HAZ): Below the hardened layer, may experience grain coarsening or tempering.

Base metal: Remains unaffected in an ideal case [56].

The key considerations for effective laser hardening include the selection of appropriate laser process parameters and the geometry of the workpiece employed.

8.5.2 LSH Process Parameters

LSH is governed by a range of laser-material interaction parameters that collectively determine the effectiveness, quality, and selectivity of the hardened surface. These parameters become particularly critical when applied to AM components due to their unique microstructural anisotropy, surface roughness, and thermal behaviour. Careful tuning of the process properties ensures sufficient hardening while preserving the geometric and structural integrity of intricate AM parts [57]. The key process parameters include laser power, scanning speed, spot diameter, and the number of passes, which collectively determine the effectiveness of the hardening process.

8.5.2.1 Laser Power (P)

Laser power refers to the energy output of the laser per unit time (in W or kW) and directly influences the peak surface temperature and the depth of the austenitised zone [58,59]. Higher power enables deeper heat penetration, allowing for thicker hardened layers. However, in AM parts, high laser power can exacerbate residual stress and thermal distortion, particularly in thin-walled or high-aspect-ratio structures [60].

8.5.2.2 Scanning Speed (v)

Scanning speed determines the interaction time between the laser and the material surface. Lower speeds allow more energy deposition, resulting in deeper hardening; conversely, higher speeds reduce the heat input and produce shallower hardened zones [61]. In AM components, maintaining a uniform scanning speed is challenging but essential for achieving consistent hardening across complex geometries. A slower scanning speed allows for more heat accumulation, enhancing hardness, while a faster speed can lead to insufficient hardening [62].

8.5.2.3 Laser Spot Size (d) and Beam Shape

The spot size, typically ranging from 0.5 to 8 mm, controls the energy density. A smaller spot concentrates energy, promoting deeper but narrower hardened zones, whereas a larger spot results in a broader yet shallower effect. The beam shape also plays a role: top-hat distributions yield uniform hardening, while Gaussian profiles result in non-uniform heating, which can be detrimental in precision AM parts [63,64].

8.5.2.4 Number of Passes

In LSH, the number of passes refers to how many times the laser beam is scanned over a specific surface area. This parameter plays a significant role in determining the final surface properties, such as hardness, microstructure, depth of the hardened layer, and residual stress profile. Single-pass treatment is generally sufficient for thin surface layers and is preferred for delicate AM geometries where thermal input must be minimal, while multi-pass strategies are often employed for thick or hardenable steels requiring deeper hardening, AM parts with variable thicknesses or complex lattice features, and enhancing coverage in areas where a single pass may leave unprocessed spots due to surface curvature or laser shadowing effects [65].

8.5.2.5 Wavelengths

The laser wavelength influences absorption efficiency, which varies by material. Shorter wavelengths (e.g. 1064 nm for YAG or fibre lasers) are absorbed more efficiently by metals, whereas longer wavelengths (e.g. 10.6 μm for CO_2 lasers) may suffer from higher reflectivity, especially on polished surfaces [66]. Aluminium and titanium alloys used in AM are highly reflective; therefore, enhancing absorptivity through surface roughening or applying absorbent coatings may be necessary.

8.5.2.6 Pulse Mode: Continuous Wave vs. Pulsed

CW lasers provide steady energy and are typically used for uniform hardening over large areas. Pulsed lasers, especially in nanosecond, picosecond, or femtosecond ranges, offer high peak power and localised hardening, ideal for intricate or thin features common in AM parts. Pulsed processing also minimises thermal diffusion, reducing the size of the HAZ [67].

8.5.2.7 Energy Density (E)

Energy density is the energy delivered per unit area (J/cm²), commonly expressed as follows:

$$E = \frac{P}{v \cdot d} \tag{8.1}$$

where P is the laser power, v is the scanning speed, and d is the spot diameter.

It determines the extent and nature of thermal transformation. An optimal energy density is necessary to reach the austenitisation temperature without surface melting or cracking. Too low a density fails to induce hardening, while too high a density can damage the surface [61].

8.5.2.8 Shielding Gas Environment

Shielding gases such as argon or nitrogen are used to prevent oxidation during laser exposure. This is especially critical for reactive AM materials like titanium and aluminium. Titanium can form brittle oxides if exposed to air at high temperatures, compromising biocompatibility and fatigue resistance [68].

8.5.3 Workpiece Geometry

Laser energy absorption is highly dependent on various factors, including surface structure and orientation. One of the most critical influences is the angle of laser incidence. In cases where components with complex geometries, such as crankshaft fillets, are subjected to laser treatment, the surfaces cannot be approximated as planar. As a result, the angle at which the laser beam strikes the surface varies locally, leading to corresponding fluctuations in energy absorption efficiency [69]. Therefore, the complexity of the part affects the choice of parameters [70].

LSH presents a highly effective and versatile post-processing technique for enhancing the surface performance of AM metallic components. By utilising controlled laser-induced heating followed by rapid self-quenching, LSH induces favourable microstructural transformations, such as martensitic phase formation or grain refinement, resulting in significant improvements in hardness, wear resistance, and fatigue performance. By integrating LSH into the AM workflow, manufacturers can unlock enhanced functionality for parts in demanding applications across aerospace, biomedical, tooling, and automotive sectors. Future directions point towards in-situ LSH during AM fabrication, Artificial Intelligence (AI)-driven parameter optimisation, and hybrid manufacturing platforms to further streamline this critical surface engineering step.

8.6 THE FLUIDISED BED FINISHING TECHNOLOGY OF AM PARTS: THIN-WALLED STRUCTURES

In the following two sections (8.6 and 8.7), the *FBF* and the *Chemical Machining* are presented as very promising *tool-less* surface finishing techniques for metal AM parts. A comprehensive overview of the fundamental mechanisms occurring during these processes is reported, according to the existing literature. A detailed description of the fundamental principles of these two processes is reported below in two sections.

FBF is framed in the wider *Abrasive Flow Machining* processes involving abrasive particles to effectively polish AM parts [71]. It is a very cost- and time-effective polishing technology, overcoming almost all the limitations of mechanical surface treatments. Moreover, the further advantages of the *Fluidised Bed*-based polishing technology are the easy process scale-up and the environmentally friendly process conditions [72].

8.6.1 Basic Principles of Fluidisation and Applications

In its essential definition, fluidisation is a process in which a solid phase gains the typical mobility of a fluid once it is invested employing a liquid or gaseous carrying flow [73]. More specifically, if the solid particles are contained in a reactor – and held in stationary conditions by a porous plate – generally called a *distributor* – the fluidising medium can penetrate between the gaps of the particles, leading to a dragging force that overcomes the gravity force of the particles or more, but without letting them to be thrown out of the reactor. The subsequent fluidised state of the solid phase possesses very intriguing properties, such as the ability to acquire the shape of the reactor itself, the liquid-like communicating vessels capability, and outstanding heat exchange ability, just to mention a few. Moreover, depending on the characteristics of the solid particles (mainly density, shape, and size) as well as the selected type of fluidising medium and flow rate, different regimes of fluidisation are possible: increasing the flow rate progressively, the regimes switch from minimum to bubbling, slugging, turbulent, and pneumatic [73], as illustrated in Figure 8.7. These regimes allow the operation through fluidisation in different types of plants, where the highest flow rate conditions (turbulent, pneumatic) require the employment of closed-circulating reactors. For these reasons, fluidised beds gained increasing attention over the years from both academia and industry for a consistent number of applications [73].

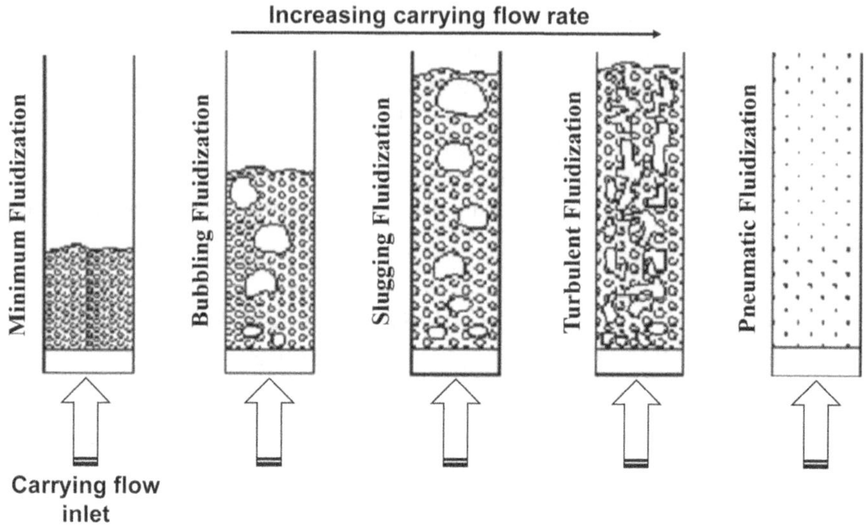

FIGURE 8.7 Fluidisation regimes as a function of carrying species flow rate.

Since the 40s, the very first industrial use of fluidised beds was for the catalytic cracking of petroleum, which represents even today the main method to produce fuels: this process is generally referred to as fluid catalytic cracking [73]. In this context, catalyst particles typically made of zeolites do represent the most important part of the cracking process. The fluidisation of the catalyst occurs through a carrying flow of preheated and long-chain hydrocarbon feedstock: once this process occurs, the proper cracking of the feedstock takes place most uniformly and efficiently, considering the combined benefits of fluidisation and catalytic activity of the particles, which must be regenerated continuously [74].

Later, an efficient and cost-effective manufacturing process for polymers such as polyethylene and polypropylene was also developed, starting in the 1970s [75]. In this case, the fluidised bed is still made of solid catalyst particles, but the fluidising species is represented by the monomers of the selected material. Afterwards, in a stable dense-phase regime (i.e. minimum fluidisation or bubbling regimes), the polymers start to grow on the surface of the catalyst particles until they reach the desired size. The so-produced material is then detached from the catalyst and automatically collected in a proper section or even directly processed to produce larger semifinished products through extrusion [75].

Today, the number of industrial applications that involve fluidised beds is remarkable. Moreover, the fact that they are also very different from each other witnesses the importance and the flexibility of use of fluidisation technology. Apart from the very first ones mentioned above, other promising uses are related to the fields of renewable energy [76], heat exchanging [77], fast pyrolysis of discharge polymers and biomasses [78,79], preparation of metal mixtures for different industrial purposes such as powder metallurgy and AM [80–83], wastewater treatment [84] and drying of bulk solids [85]. This list, which is not exhaustive, is even more intriguing considering the increasingly demanding requirements for sustainable and efficient processing of materials and products.

8.6.2 The Role of Fluidised Beds in the Surface Processing Industry

The very first use of fluidised beds for surface processing was for coating metal parts by means of polymers. If the latter, in the form of particles, are fluidised at a temperature close to the melting point of the polymer itself, then the coating of the metal can be performed by dipping it in the reactor for a few to tens of seconds [73]. Afterwards, since the deposition thickness as a function of time does not follow a linear relationship, the final thickness and smoothness of the coating can be controlled by reheating the metal in hot air. This process has been a research subject matter over the last years and has also found application in the pharmaceutical and food sectors [86,87]. However, it is worth mentioning that in these applications, the object of coating was the fluidised particles themselves rather than an external part introduced in the reactor: this process is generally recalled in literature as *top-spray fluidised bed coating* [88], and it involves an additional inlet in the reactor, generally placed near to the distributor plate, which has the function of introducing the coating species.

In turn, the use of the fluidised bed technology as a surface finishing method has a relatively recent origin, and it is still a matter of research study rather than a mature process today. In this context of use, the fluidised solid particles play the role of the finishing medium once a part is somehow immersed in the bed. This finishing solution, compared to the other ones actually available, presents several potential benefits that can be synthesised as follows: (i) in comparison with the more traditional grinding-based methods like vibratory polishing, sand blasting and SP, the use of fluidised beds could guarantee more flexibility in terms of processing complex surfaces, accessing also inner features that other processes are incapable of, combined with a robust control of the finishing parameters and homogeneity of the results; (ii) compared to CNC machining or laser polishing, the previous point is even more enforced, without the need of expensive equipment; (iii) compared to the processes based on the chemical and electrochemical polishing principles, which represent probably the most appealing solutions for complex parts, the fluidised bed-based finishing has no practical environmental impact since harmless substances are used; (iv) the ease of construction and operation of the finishing plant combined with the low operational costs and high surface quality achievable make the FBF very competitive to any other finishing method. According to several experimental works reported in the literature, the surface finishing process based on the fluidised bed is referred to as *Fluidised Bed Machining*, *Fluidised Bed-Abrasive Jet Machining*, or *Abrasive Fluidised Bed (AFB)*. For the sake of clarity, this process will be denoted from now on as *FBF*.

The first systematic and documented research activity aimed at exploring the potentialities of FBF was carried out by Barletta [89]. In his work, the effects of the main process variables on the surface quality and wear evolution of AA2024 flat substrates have been analysed, namely the finishing time, the characteristics of the abrasives (material and size), and the air flow rate used for the fluidisation. From the point of view of the substrate, the experimental work also included the investigation of different starting roughness values (R_a: 0.2 ÷ 1 μm) and heat treatment conditions (O,

T3). The results proved the viability of the process both in terms of surface quality improvement and limited alteration of the dimensional accuracy, highlighting also the primary importance of the starting roughness on the final result. More specifically, one of the important experimental outcomes that emerged was that an asymptotic value of roughness is reached once the process conditions are set. Accordingly, FBF might be beneficial or not, depending on the case study considered. Moreover, the FBF feasibility was also evaluated for more complex geometries, proving its validity again.

In a later research, Barletta et al. investigated an alternative FBF configuration designed for the processing of tubular structures characterised by a high length-to-diameter ratio [90]. More specifically, the process configuration essentially consisted of two communicating fluidised beds working alternatively, between which the tubes to be finished were inserted. The process demonstrated again to be efficient, and more highlights concerning the influence of the process variables were provided, namely the strong correlation between the reduced size of the abrasive particles and the improvement of the surface quality. The same experimental apparatus was used by Barletta et al. in their work concerning the finishing of Inconel 718 alloy tubes [90]. Using alumina abrasive particles, they demonstrated that the finishing operation was feasible also for harder substrates compared to aluminium alloy ones, but still preserving the characteristic of a non-pressure copying technique and also improving the cylindricity tolerance of the parts.

A more comprehensive description of the main FBF process features, benefits, and drawbacks, as well as its applications and dynamics, was provided later in a review article by Barletta [91]. This work, in the first place, condensed the description of the main phenomena involved in the finishing of objects through fluidised beds. Concerning the specific case of metals, the review highlighted the fundamental interaction mechanisms between the fluidised particles and the substrates, namely micro-ploughing, micro-peening, micro-cracking, and micro-fatigue. These phenomena, whose occurrence depends on the main relative motion of the particles on the surface (i.e. rolling and/or sliding), were first presented and analysed earlier in the more general framework of the erosive models of Finnie [92], Khruschov and Bavichev [93], and Gahr [94]. Therefore, depending on the FBF operating parameters, such as fluidisation regime, abrasive type, processing time, angle of the substrate experiencing the finishing, and the process configuration, one may expect a notable variability in the achievable results. The review work also included further applications of the fluidised beds for the surface processing, apart from the strict finishing, such as the degreasing of parts through a fluidised bed made of glass beads and the roughening of the surfaces of machining tools before coating through the chemical vapour deposition process. However, the list does not pretend to be exhaustive, as the potential applications of surface processing through fluidised beds are surely greater and rapidly evolving.

8.6.3 State of the Art, Current Challenges, and Future Perspectives for the Finishing of Metal AM Parts

The experimental work concerning the FBF mentioned in the previous section highlighted, amongst the several process variables, the importance of the starting surface quality of the parts on the finishing capability of AFBs. However, most of the experimental work aimed at exploring the capabilities of FBF involved substrates and parts produced according to traditional manufacturing methods. For instance, it is well known that the common manufacturing processes, like CNC machining, sheet forming, and bulk forming, generally lead to very high surface quality, thus not requiring intensive post-process finishing steps, if any. On the other hand, the scenario changes completely when the parts are produced through AM, especially considering the cases of metal parts made by powder-based processes, which represent the worst case in this framework and thus, become the focus of the following discussion. In addition, most of the erosion theories and models applied to the case of FBF become less reliable, as additional phenomena involved in the interaction between rough surfaces and fluidised particles are neglected.

According to the description of the phenomena that lead to the poor surface quality of the powder-based metal AM parts, the resulting texture of the latter is very rough and characterised by a random pattern of sintered powders. Accordingly, as the surface roughness of as-built parts (mostly quantified through Ra and Sa) lies in the range between 5 and 50 μm – a range that includes variable metal AM processes, parameters, and materials [95] – a different response to FBF is expected in comparison with the smoother surfaces produced with the more conventional manufacturing routes. In this framework, Atzeni et al. were the first to experimentally investigate the ability of FBF to perform the surface finishing of metal AM parts, considering the case study of AlSi10Mg flat substrates produced through Laser-Powder Bed Fusion (L-PBF) [96]. The finishing process, however, was investigated in a custom configuration: the specimens were put into rotation at relatively high speed after the immersion in a reactor containing steel abrasive particles at the minimum fluidisation regime. In this way, a relative motion between the exposed surface and the fluidised bed is ensured without the need for turbulent fluidisation regimes. This choice guarantees a more uniform smoothing effect compared to the case of the impact of the abrasives onto surfaces placed in a static way in the reactor. The experimental study allowed, within the investigated conditions, us to conclude that FBF was viable and significant improvements of the surface quality were achieved (considering in the first place only Ra, the roughness was reduced by more than 90% of the as-built value). Following a similar concept, Atzeni et al. also investigated the capability of FBF to improve the surface quality of Ti6Al4V specimens produced through electron beam powder bed fusion (E-PBF) [97]. As is well known, E-PBF is capable of higher productivity in comparison to L-PBF due to the higher power of the electron beam against the laser, leading to the capability of processing powder feedstocks with greater sizes [98]. In turn, this aspect leads to the drawback of poorer surface quality, therefore representing a more challenging case for the FBF. However, by selecting high rotational speeds and testing longer process times, the FBF process demonstrated its validity for this case as well, achieving very similar results to the case of AlSi10Mg presented in [96] in terms of percentage roughness reduction. A more comprehensive experimental work, aimed at exploiting the influence of the operational parameters on the surface finishing ability of FBF, was carried out by Troiano et al. [99]. Despite the experiments being focused only on flat AlSi10Mg alloy specimens produced via L-PBF, similarly to the work of Atzeni et al. [96], the presented data allowed for an enlargement of the overview of the influence of finishing conditions in terms of abrasive material and process configuration. More specifically, the experiments were aimed at analysing the effects on the surface texture evolution and wear for four abrasive particles, classified according to their "maximized" property: silica sand and corundum were selected according to their low density and high hardness, two choices that would guarantee a lower air flow rate for the fluidisation and high wear compared to steel particles, respectively. The latter, on the other hand, maximises the energy transfer towards the surfaces to be smoothed, due to the higher density compared to the previous case, but still preserving a sufficient hardness level. Moreover, for the steel particles, the influence of their shape was included by considering grit and cut-wire feedstocks. Concerning the process configuration, two experimental setups were considered: (i) static positioning of the specimens and bubbling fluidisation regime and (ii) rotating specimens immersed in an abrasive bed kept at the minimum fluidisation regime. In both cases, the tilt angle of the specimens was considered as a process variable as well. The results obtained from the experiments allowed us to draw some interesting conclusions about the smoothing dynamics and how to improve the finishing capability of FBF when rough surfaces are considered. To summarise, the following experimental outcomes were enucleated: (i) the density of the abrasives plays a primary role in the finishing process. The higher the density, the greater the surface modification; (ii) the rotating-specimen configuration proved to be more viable in comparison with the static specimen one, a conclusion that was in agreement with the results obtained by Atzeni et al. [96]; (iii) the relative tilt angle between the specimen and the abrasive should maximise the shear stresses rather than the compressive ones. The latter are typically generated in cases of impact of the abrasives onto the surface, considering, for instance, the conditions of the static specimen experiments. Another interesting outcome that emerged from the experiments was that the aforementioned findings were in agreement with the Bagnold equation that models the shear stresses exerted by fluidised abrasive particles onto plane surfaces as a function of characteristics of the latter and the relative speed [100]. More specifically, the agreement between the experiments and the shear model lies in the dependence of the surface modification ability of the fluidised particles on their density and relative speed, with a linear and quadratic power, respectively.

All the experimental works conducted in the field of FBF of metal parts made by AM proved the validity of this process to perform the finishing also on very rough parts. However, some of the actual limitations and challenges for further process development concern the ability to finish complex parts such as lattice structures: this capability is also a function of the proper choice of the process configuration and abrasive characteristics, as witnessed by the results mentioned in this section. Moreover, the use of a rotating-part FBF configuration requires a proper finishing strategy to be developed to guarantee the desired uniformity of the surfaces [99]. Some promising approaches of novel FBF configurations, which represent today an unexplored research field, might involve the use of different external force fields, such as magnetic or mechano-chemical ones, in substitution for the additional rotation. These options would be helpful to gain the benefits of the non-directional nature of the fluidisation, combined with the capability of the process to obtain satisfactory results of surface quality in a cost- and time-effective manner.

8.7 CHEMICAL MACHINING OF AM PARTS: THIN-WALLED STRUCTURES

The *Chemical Machining* emerges as a particularly compelling solution, especially when considering complex geometries. Their primary advantage lies in the ability of the chemical HIP post-processing technique, which was developed in the mid-1950s, to reduce porosities and further densify the consolidated powders of sintered metals, ceramics, or composite material parts within high-temperature and isostatic high-pressure environments [23]. The HIP process involves the placement of the as-built AM part in a sealed container with inert gas. This sealed container is then positioned within a high-temperature (up to 2000°C) and high-pressure (up to 2700°C) vessel (see Figure 8.3). The processing temperature of HIP is usually below the melting point and higher than that of the solidus temperature line, as per ASTM F3301 [23]. During the HIP process, the as-built AM samples are subjected to mechanical pressing under isostatic pressure and elevated temperatures, and aggressive solutions are used to access the internal features of complex parts, such as internal channels and *lattice structures.* This accessibility sets them apart from the more commonly used mechanical and thermal treatments, and positions chemical methods as having significantly greater application potential [101]. Numerous researchers have conducted experimental studies on chemical polishing [102,103]. For example, Tyagi et al. [104] utilised chemical treatments to polish SS316L components, achieving highly satisfactory results with surface roughness (Ra) reductions of approximately 92% on both internal and external surfaces. Similarly, it is important to acknowledge the limitations associated with chemical smoothing processes. These include the use of toxic and environmentally harmful chemical baths, challenges in achieving precise dimensional control, high operational costs, and other related issues.

Chemical machining is a widely adopted technology for processing metallic materials. This subtractive process relies on the metal's reactivity and subsequent dissolution in specific environments, typically involving acidic or alkaline aqueous solutions. Commonly used in biomedical and aerospace industries, chemical machining offers distinct advantages, such as avoiding the creation of altered zones on the processed components. Beyond material removal, it can enhance surface finish by tailoring the solution's composition.

A key benefit of chemical machining is its ability to handle complex geometries without requiring specialised tools. This capability aligns well with the advantages of AM, which excels in producing intricate shapes.

With appropriate chemical solutions, various metal alloys can be effectively treated. For instance, steels and copper alloys are processed using ferric chloride ($FeCl_3$) solutions, while titanium and silicon-containing aluminium alloys are treated with mixtures of nitric and hydrofluoric acids. Despite its advantages, chemical machining presents challenges, such as the environmentally harmful disposal of spent solutions. Moreover, unlike traditional machining, where removed material (e.g. chips) can often be recycled, the dissolved material in chemical machining cannot be recovered.

8.7.1 Reactions in Chemical Machining

Chemical machining is a technology that allows material to be removed by exploiting its reactivity to a specific substance.

The process is made in aqueous solution, typically acidic or alkaline. For instance, in the case of an acidic solution, the material removal is based upon the following reaction:

$$M \rightarrow M^{+} + e^{-} \tag{8.2}$$

$$H^{+} + e^{-} \rightarrow \frac{1}{2} H_2 \tag{8.3}$$

It is substantially a corrosion process where the energy given by the cathodic reduction of hydrogen allows the dissolution of metal. Obviously, the M^{+} ion must be fully soluble in the solution, and in the case of an alkaline environment, the cathodic reaction is:

$$H_2O + e^{-} \rightarrow \frac{1}{2} H_2 + OH^{-} \tag{8.4}$$

For copper, nickel alloys, and steels, the solutions of ferric chloride ($FeCl_3$) are widely used; the reduction of ferric ions (Fe^{3+}) to ferrous ions (Fe^{2+}) is the energy-giving cathodic process.

$$Fe^{3+} + e^{-} \rightarrow Fe^{2+} \tag{8.5}$$

For alloys that easily passivate, like titanium and aluminium, the chemical machining process often involves the formation of metal oxide; the reaction network should be very complex; however, in the following, it is schematically reported for a monovalent metal ion and the cathodic process (8.2):

$$M \rightarrow M^{+} + e^{-} \tag{8.6}$$

$$2M^{+} + H_2O \rightarrow M_2O + 2H^{+} \tag{8.7}$$

In order for the process to proceed, the presence in the solution of a molecule able to dissolve the oxide is necessary:

$$M_2O + A \rightarrow M_2O \cdot A \tag{8.8}$$

A is a generic chemical species capable of forming soluble compounds with metal oxides; for example, fluoride anions can dissolve both titanium and aluminium oxides.

In the previously described reaction network, the dissolution of metal occurs via reactions (8.6), (8.7), and (8.8), and the rate-limiting step of the process is the oxide

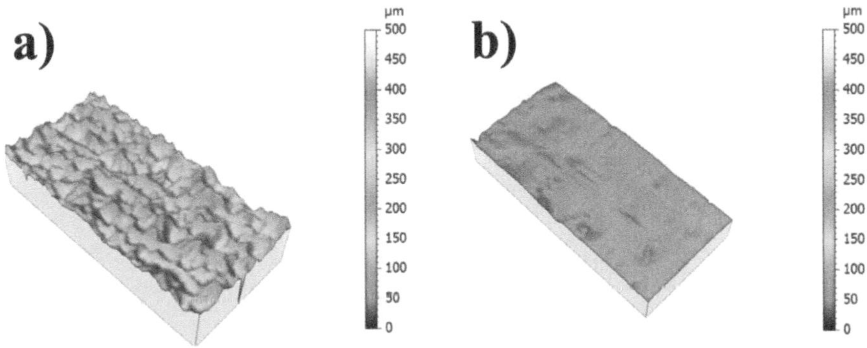

FIGURE 8.8 Morphology of the surface of titanium Grade 5 treated in HF solution (a) and in HF-HNO_3 solution (b, HF/HNO_3 molar ratio equal to 0.3) [105].

formation through reaction with water. In fact, water is a weak oxidising agent; for this purpose, a strong oxidising agent, like nitric or sulphuric acid, is added. Another challenge in chemical machining is the formation of hydrogen gas. This issue poses multiple risks: not only is hydrogen highly explosive, creating safety concerns, but it can also lead to hydrogen embrittlement, compromising the material's structural integrity and causing irregularities in the machining process. Hydrogen produces gas bubbles on the metal surface, which prevent contact between the metal and the solution. As a result, the process does not proceed uniformly across the entire surface; to this end, nitric acid can avoid the formation of hydrogen bubbles.

The effect of hydrogen formation on the morphology of a Grade 5 titanium surface treated with an aqueous solution of hydrofluoric acid is shown in Figure 8.8, where, for comparison, the surface of the same material treated in a solution also containing nitric acid is also shown. It is observed that the addition of nitric acid, which inhibits gas formation on the surface, significantly improves the surface finish.

Two possible reaction networks for the chemical machining of metallic materials have been illustrated. The former involves the direct dissolution of the metal, while the latter involves dissolution through a process of passivation and de-passivation. Only in the second case were the conditions for brightening achieved, resulting in a shiny surface.

8.7.1.1 Example: Reaction in Chemical Machining of Titanium Alloys in HF-HNO_3 Solutions

According to Sutter and Goetz-Grandmont [106] and Sutter et al. [107], in aqueous solution of hydrofluoric acid, the dissolution of titanium occurs through the following chemical reactions:

$$Ti \rightarrow Ti^{3+} + 3e^- \tag{8.9}$$

$$Ti^{3+} + 6HF \rightarrow [TiF_6]^{2-} + 6H^+ + e^- \tag{8.10}$$

$$4H^+ + 4e^- \rightarrow 2H_2 \tag{8.11}$$

Since the complex molecule $[TiF_6]^{2-}$ is soluble, there is material removal. The process based upon HF is used, for example, to treat prosthetics in the biomedical industry; the metal surface results in rough and irregular surfaces due to gas bubble formation; however, this kind of surface is ideal for osseointegration.

To obtain a more regular and smoother surface, nitric acid is added, and the reaction network is modified:

$$Ti^{3+} + \frac{1}{2}HNO_3 + \frac{3}{2}H_2O \rightarrow TiO_2 + \frac{1}{2}HNO_2 + 3H^+ \quad (8.12)$$

$$TiO_2 + 6HF \rightarrow [TiF_6]^{2-} + 2H_2O + 2H^+ \quad (8.13)$$

In addition to its strong oxidising action, nitric acid inhibits the formation of hydrogen on metal surfaces:

$$3H_2 + HNO_3 \rightarrow 4H_2O + 2NO \quad (8.14)$$

In summary, in hydrofluoric acid solution, the dissolution of titanium occurs through reactions (8.9) and (8.10) in the presence of nitric acid; an oxidising step (reaction 8.12) is added, and the dissolution takes place via reactions (8.9), (8.12), and (8.13).

Some additional considerations must be made when dealing with alloys other than commercially pure titanium. One of the most used titanium alloys is Grade 5 titanium, which contains 6% aluminium and 4% vanadium as alloying elements. This alloy belongs to the alpha + beta category because its microstructure consists of alpha phase (hexagonal close-packed structure) and beta phase (body-centred cubic structure). Additionally, this alloy is amongst the most used in AM processes.

The reactivity of the alpha and beta phases towards the nitric and hydrofluoric acid solutions is different. The process, as previously mentioned, is a highly accelerated corrosion with anodic and cathodic reactions. In Grade 5 titanium, aluminium is selectively dissolved from the alpha phase, while vanadium is dissolved from the beta phase. This indicates that the beta phase, in terms of corrosion behaviour, exhibits greater nobility and predominantly acts as a cathodic region. As a result, the alpha phase undergoes preferential dissolution.

Figure 8.9 shows the micrographs of a Grade 5 titanium component produced by electron beam melting. The initial microstructure (Figure 8.9(a)) is typical of this process, with alpha-phase lamellae separated by beta phase. Comparing the same microstructure after treatment in a nitric and hydrofluoric acid solution (Figure 8.9(b)), it can be observed that the beta phase is slightly raised, confirming the previously mentioned selectivity of the process.

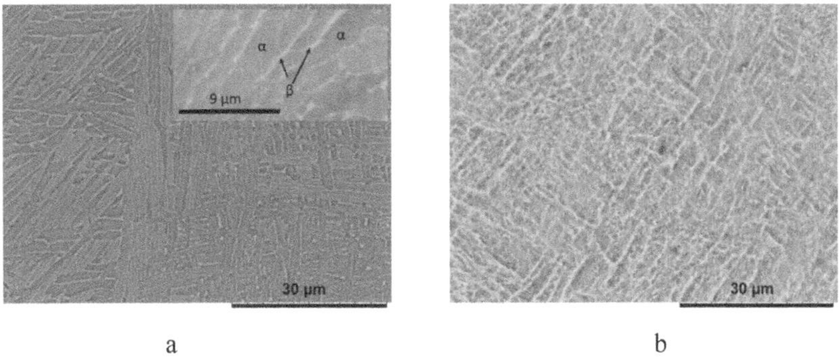

FIGURE 8.9 Microstructure of as-built titanium Grade 5 component made by electron beam melting with highlighted alpha and beta phases (a); the same component after treatment in HF-HNO_3 solutions, the beta phase appears in relief with respect to the alpha phase (b) [105].

8.7.2 The Smoothing Effect

The interest in using chemical machining to treat components made through additive technology lies in the possibility of obtaining smooth surfaces with few defects. To understand the conditions that lead to a smoothing effect, it is necessary to make some kinetic considerations about the process.

Chemical machining is a process that involves a heterogeneous chemical reaction between a metal surface and a solution. The kinetics of the process consists of two steps in series: the former is the chemical reaction on the metal surface, and the latter is the mass transport of products and reactants between the surface and the bulk of the solution. On a rough surface, ideally consisting of alternating peaks and valleys, the smoothing effect is achieved only if the process is faster on the peaks than in the valleys.

The following is a simplified discussion of the main kinetic aspects of a chemical machining process, as reported by Kim and Suh [108]. In Cartesian coordinates, the equation describing the distribution of a chemical species inside a solution is:

$$\frac{\partial C}{\partial t}+u\frac{\partial C}{\partial x}+v\frac{\partial C}{\partial y}+w\frac{\partial C}{\partial z}=D\left(\frac{\partial^2 C}{\partial x^2}+\frac{\partial^2 C}{\partial y^2}+\frac{\partial^2 C}{\partial z^2}\right) \quad (8.15)$$

where C is the concentration; t is time; x, y, and z are the spatial coordinates; u, v, and w are the three components of fluid velocity; and D is the diffusivity.

If the metal surface is stationary, then the fluid velocity near the surface is zero. Let (d) be the thickness of the diffusive boundary layer, which is the distance from the surface where the concentration changes from the surface value C_0 to the bulk value C_b in the solution. Assuming that concentration variations occur only in the direction normal to the surface (y), then equation (8.15) can be simplified to:

$$\frac{\partial^2 C}{\partial y^2} = 0 \tag{8.16}$$

This equation is valid in steady-state conditions and can be solved with the following boundary conditions:

$$C = C_0 \text{ at } y = 0$$

$$C = C_b \text{ at } y = \delta$$

The solution is:

$$C = \frac{C_b - C_0}{C_0} y - C_0 \tag{8.17}$$

In steady-state conditions, the same amount of material reacting on the surface is transported to the bulk of the solution, which means that the diffusive flux on the surface is equal to the reaction rate:

$$DD\left(\frac{\partial C}{\partial y}\right)_{y=0} = kC_0 \tag{8.18}$$

where k is the reaction constant, combining equation (8.17) with condition (8.18) yields:

$$\frac{D}{k\delta} = \frac{C_0}{C_b - C_0} \tag{8.19}$$

If $\frac{D}{k\delta}$ approaches zero, and then the term $\frac{C_0}{C_b - C_0}$ also approaches zero, which means that the concentration on the surface is almost zero. This condition occurs if the reactants are immediately consumed as soon as they reach the surface, indicating that the process is diffusion-controlled, meaning the reaction rate is much higher than the diffusive flux.

This phenomenon is commonly observed in chemical machining processes where the reactions between the metal and the solution occur rapidly. However, for the smoothing effect to take place, it is not sufficient for the process kinetics to be merely diffusion-controlled. Ideally, on a surface with a rough profile, if the reaction rate is uniform across all points, the treatment would simply remove material without altering the profile's shape.

If the diffusion boundary layer is significantly thicker than the average height of the rough profile, the smoothing effect will not occur. For smoothing to take place,

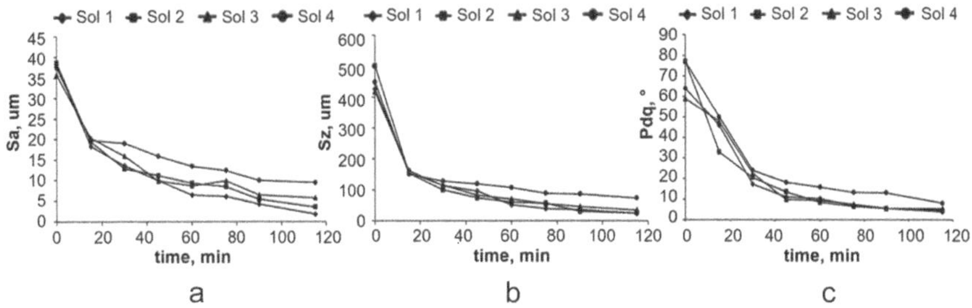

FIGURE 8.10 Trend of S_a (a), S_z (b), and P_{dq} (c) for titanium Grade 5 components realised by EBM in HF-HNO_3 solution with different molar ratios (Sol 1 HF/HNO_3 = 0.3, Sol 2 HF/HNO_3 = 0.6, Sol 1 HF/HNO_3 = 1, Sol 4 only HF) [105].

the diffusion boundary layer must be comparable to or smaller than the profile's average height. While this condition may be present at the start of the process when the surface roughness is high, it often ceases to exist as the treatment progresses, resulting in the cessation of smoothing and continuation of uniform material removal.

Chemical treatment is particularly effective for surfaces with roughness profiles characterised by sharp peaks and narrow, deep valleys, such as those found on components produced via powder bed AM. In such cases, the transport of reactants and by-products within the valleys is further hindered by geometric constraints, making the selective dissolution of peaks more pronounced. Conversely, surfaces with wavy profiles are treated less effectively.

The progression of roughness during a chemical smoothing process applied to a component manufactured using powder bed additive techniques is depicted in Figure 8.10. The trend is asymptotic, with significant smoothing occurring within the first 30 minutes of treatment. Beyond this point, the roughness stabilises and remains constant.

In Figure 8.10, it is interesting to compare the trends of S_a and S_z with those of P_{dq}, which represents the average slope of the roughness profile. Initially, when the profile is sharp and characterised by steep slopes, the smoothing effect is highly pronounced, leading to a significant reduction in roughness. However, as the profile becomes more rounded and P_{dq} decreases, the selective dissolution of the peaks diminishes. Consequently, the rate of roughness reduction slows, and the roughness stabilises, tending to remain constant over time.

8.7.2.1 Example: Smoothing and Brightening of the AlSi10Mg Component Made via Laser-Powder Bed Additive Manufacturing

Chemical machining is a process commonly applied to aluminium alloys. Ferric chloride solutions are typically used, and machining is followed by a brightening step to give the material the desired finish.

The AlSi10Mg alloy used in additive processes contains 10% silicon, and ferric chloride is unable to dissolve this element. The AlSi10Mg samples were treated in two steps: the first step involved chemical machining using HF-HNO_3 solutions to remove irregularities such as unfused powders present on the surface, and the second

TABLE 8.1
Compositions of employed solutions used in chemical machining (CM) and chemical brightening (CB) treatments

CM		CB	
HNO_3 (69.5 wt %)	50 ml	HNO_3 (69.5 wt %)	58 ml
HF (50 wt %)	20 ml	HF (50 wt %)	58 ml
H_20	930 ml	H_3PO_4 (85 wt %)	720 ml
		H_2SO_4 (96 wt %)	164 ml
		$CuSO_4$	0.5 g

step involved chemical brightening with phosphoric acid solutions to give the final finish. The composition of the employed solution is reported in Table 8.1.

The chemical reactions that occur during the treatment with *HF-HNO*$_3$ are simplified below:

$$2Al+6HNO_3 \rightarrow Al_2O_3+3H_2O \tag{8.20}$$

$$Si+HNO_3 \rightarrow SiO_2+4NO_2+2H_2O \tag{8.21}$$

and

$$Al_2O_3+6HF \rightarrow 2AlF_3+3H_2O \tag{8.22}$$

$$SiO_2+4HF \rightarrow SiF_4+2H_2O \tag{8.23}$$

Since AlF_3 and SiF_4 are soluble, material removal occurs. The process proceeds via passivation and de-passivation in the same manner as titanium.

Upon completion of the chemical machining treatment, a notable reduction in surface defects was observed, particularly in the unmelted powder, as illustrated in Figure 8.11. Nevertheless, a distinct etching effect is discernible, with the edges of the melting pools discernible on the sample surface. It is important to note that chemical treatments are highly susceptible to surface non-uniformities, both in terms of composition and microstructure.

In the brightening solution, the process occurs via a two-stage mechanism involving passivation and subsequent de-passivation. The formation of an oxide layer, comprising nitric and sulphuric acids, is facilitated by the action of phosphoric and hydrofluoric acids, which subsequently dissolve the layer. The solution is based on phosphoric acid and has a high viscosity, which results in a process dependent on diffusion and the transport of matter between the metal surface and the bulk of the solution. Upon completion of the brightening treatment, the melting pools are no longer visible, and the metal surface appears uniform (see Figure 8.11). Despite the differing compositions between the bulk and the edges of the melting pools, resulting

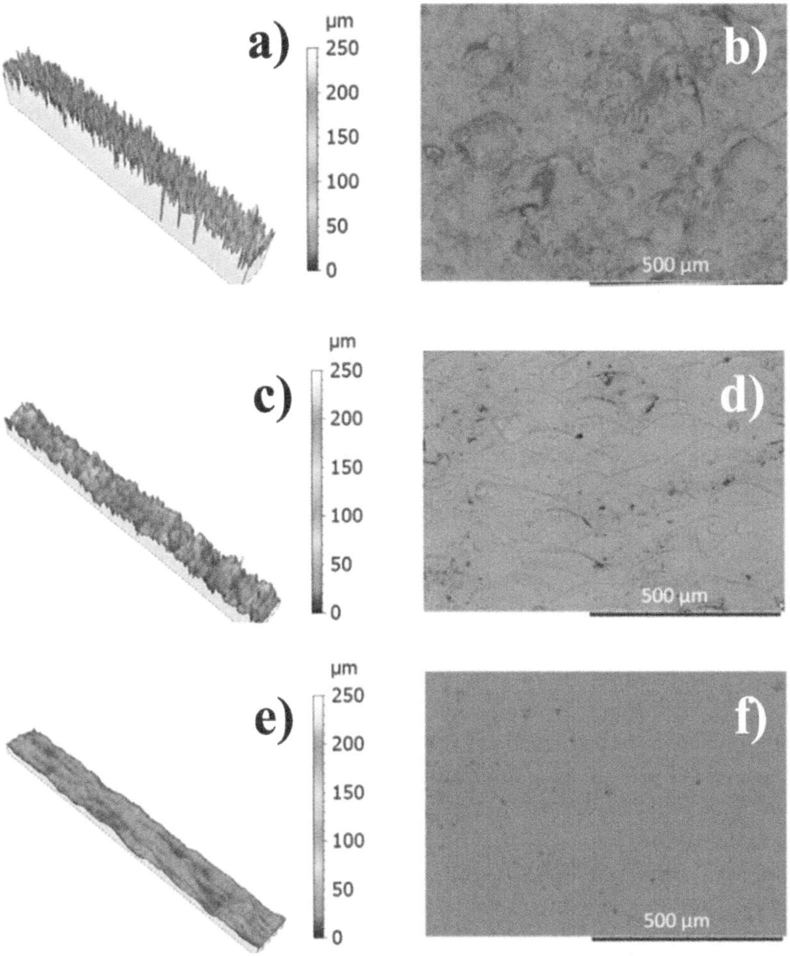

FIGURE 8.11 Surface morphology (a) and micrography (b) of as-built AlSi10Mg made by powder bed fusion additive manufacturing, after chemical machining treatment (c, d) and chemical brightening (e, f) [109].

in disparate surface reaction rates, the overall chemical brightening process is constrained by the transport of matter between the surface and the bulk of the solution. This rate is consistent across the entire surface, resulting in uniform dissolution.

One of the components of chemical brightening solutions for aluminium alloys is a noble metal salt, such as copper sulphate, in the present study. According to the author [109], the following reaction occurs during the process:

$$3Cu^{2+} + 2Al \rightarrow 3Cu + 2Al^{3+} \tag{8.24}$$

This forms a thin layer of copper on the metal surface. This layer, under the action of the acids of the brightening solution, dissolves. However, due to the diffusion reasons previously mentioned, the dissolution is greater at the peaks than in the valleys, where the copper accumulates. Thus, a galvanic couple is established between the copper present in the valleys and the aluminium peaks, which further contributes to the selective dissolution of the peaks and therefore to the smoothing effect.

8.8 SUMMARY AND OUTLOOK

In order to improve the process efficiency and ensure that the final part quality meets the specifications of industrial standards, AM processes not only need to optimise process parameters and control the quality continuously but also need to develop and adapt efficient and advanced post-processing techniques. As discussed throughout the chapter, the complex AM thermophysical phenomena induce anisotropic microstructures and intricate surface topographical fingerprints comprising various inevitable defects and asperities, which eventually affect the surface finish and mechanical performance of as-built AM parts. To minimise or tailor the anisotropic textures and to achieve the desired or acceptable surface finish or surface roughness leading to overall improved mechanical performance, one or a combination of mainstream or advanced post-processing techniques and hybrid post-processing techniques is employed. Suitable mainstream, advanced, or hybrid post-processing techniques assist in fine-tuning the as-built AM parts quality, including dimensional tolerances, and ensure that they fulfil the functional specifications of various industrial sector applications.

The prospects of advanced post-processing techniques adapted and developed to improve the overall part quality of as-built AM parts are briefly summarised: (i) developing and analysing various analytical and simulation models, which provide valuable insights into the complex thermophysical phenomena of the AM process, necessary post-processing techniques, parameter optimisation, and the potential problems. The numerical simulation models also help to study the process-structure-property interrelationship mechanisms under different conditions and act as a design guideline for experimentation and actual production, (ii) post-processing automation helps rectify human errors by minimising manual intervention, completing the post-processing tasks (cleaning, burr removing, curing, surface enhancements, etc.) quickly and efficiently. Automated systems are flexible, adopting the demands of production, thereby shortening processing times while improving the production safety and significantly reducing the labour costs, (iii) utilising the benefits offered by advanced artificial intelligence and machine learning language tools significantly helps to control, improve efficiency, and enhance the processing quality of advanced post-processing techniques, (iv) developing hybrid systems not only improves the as-built AM parts quality but also significantly improves the overall production quality of AM parts itself. For example, a hybrid direct energy deposition machine combined with advanced machining techniques greatly improves the AM production rate as well as the surface quality of as-built AM parts, and (v) developing efficient, sustainable post-processing techniques that are less hazardous to the environment without compromising post-processing quality and efficiency. This summary and outlook are completely applicable to AM thin-walled structures.

REFERENCES

1. ASTM ISO 52900-21 Additive Manufacturing—General Principles—Fundamentals and Vocabulary 2021.
2. Narasimharaju, S. R. et al. (2024). Surface texture characterisation of longitudinal and latitudinal external and internal surfaces of laser powder bed fusion processed bespoke ball artefact. *Journal of Manufacturing Processes*, 116, 120–148.
3. Matias, E., & Rao, B. (2015, August). 3D printing: On its historical evolution and the implications for business. In *2015 Portland International Conference on Management of Engineering and Technology (PICMET)* (pp. 551–558). IEEE.
4. Behera, M. P., Dougherty, T., & Singamneni, S. (2019). Conventional and additive manufacturing with metal matrix composites: A perspective. *Procedia Manufacturing*, 30, 159166.
5. Khodabakhshi, F., Farshidianfar, M. H., Bakhshivash, S., Gerlich, A. P., & Khajepour, A. (2019). Dissimilar metals deposition by directed energy based on powder-fed laser additive manufacturing. *Journal of Manufacturing Processes*, 43, 83–97.
6. Abe, T., & Sasahara, H. (2016). Dissimilar metal deposition with a stainless steel and nickel-based alloy using wire and arc-based additive manufacturing. *Precision Engineering*, 45, 387–395.
7. Gibson, I., Rosen, D., Stucker, B., & Khorasani, M. (2014). *Additive manufacturing technologies* (Vol. 17, p. 195). New York: Springer.
8. Zluhan, B., Narasimharaju, S. R., Cholkar, A., Thomas, K., Raghavendra, R., & Lopes, E. S. (2025). Design, defect analysis, compressive strength and surface texture characterization of Laser Powder Bed Fusion Processed Ti6Al4V Lattice structures. *Journal of Materials Research and Technology*, 35, 2914–2933.
9. Narasimharaju, S. R. et al. (2021). Surface texture characterization of metal selective laser melted part with varying surface inclinations. *Journal of Tribology*, 143(5), 051106.
10. Hassanin, H., ElShaer, A., Benhadj-Djilali, R., Modica, F., & Fassi, I. (2018). Surface finish improvement of additive manufactured metal parts. In Gupta, K. (Eds.), *Micro and precision manufacturing (engineering materials)*. Springer, Cham (pp. 145–164).
11. Kokkonen, P. et al. (2016). *Design guide for additive manufacturing of metal components by SLM process*, VTT Technical Research Center of Finland, Volume: VTT-R-03160-16, 1–131.
12. Fox, J. C., Moylan, S. P., & Lane, B. M. (2016). Effect of process parameters on the surface roughness of overhanging structures in laser powder bed fusion additive manufacturing. *Procedia CIRP*, 45, 131–134.
13. Delgado, J., Ciurana, J., & Rodríguez, C. A. (2012). Influence of process parameters on part quality and mechanical properties for DMLS and SLM with iron-based materials. *The International Journal of Advanced Manufacturing Technology*, 60(5–8), 601–610.
14. Miyoshi, K. Chapter 16: Surface analysis and tools. In Vižintin, J., Kalin, M., Dohda, K., & Jahanmir, S. (Eds.), *Tribology of mechanical systems: A guide to present and future technologies*. ASME Press, 2004, 293–312.
15. Leach, R. (Ed.). (2011). *Optical measurement of surface topography* (Vol. 8). Springer Berlin Heidelberg.
16. Diaz, A. (2019). Surface texture characterization and optimization of metal additive manufacturing produced components for aero applications. In *AM for the aerospace industry* (pp. 341–374). Elsevier.
17. Maamoun, A. H., Xue, Y. F., Elbestawi, M. A., & Veldhuis, S. C. (2018). Effect of selective laser melting process parameters on the quality of al alloy parts: Powder characterization, density, surface roughness, and dimensional accuracy. *Materials*, 11(12), 2343.

18. Nagarajan, B., Hu, Z., Song, X., Zhai, W., & Wei, J. (2019). Development of micro selective laser melting: The state of the art and future perspectives. *Engineering*, 5(4), 702–720.
19. Narasimharaju, S. R. et al. (2022). A comprehensive review on laser powder bed fusion of steels: Processing, microstructure, defects and control methods, mechanical properties, current challenges and future trends. *Journal of Manufacturing Processes*, 75, 375–414.
20. Huang, Y., Chen, Z. W., Wan, A. R. O., Schmidt, K., Sefont, P., & Singamneni, S. (2024). Electron beam powder bed fusion additive manufacturing of Ti6Al4V alloy lattice structures: Orientation-dependent compressive strength and fracture behaviour. *The International Journal of Advanced Manufacturing Technology*, 132(7), 3299–3311.
21. Liu, Y. et al. (2024). Post-treatment technologies for high-speed additive manufacturing: Status, challenge and tendency. *Journal of Materials Research and Technology*, 30, 1057–1082.
22. Pourrahimi, S., & Hof, L. A. (2024). On the post-processing of complex additive manufactured metallic parts: A review. *Advanced Engineering Materials*, 26(10), 2301511.
23. Ge, J., Pillay, S., & Ning, H. (2023). Post-process treatments for additive-manufactured metallic structures: A comprehensive review. *Journal of Materials Engineering and Performance*, 32(16), 7073–7122.
24. Ravi, G. A., Hao, X., Wain, N., Wu, X., & Attallah, M. M. (2013). Direct laser fabrication of three-dimensional components using SC420 stainless steel. *Materials & Design*, 47, 731–736.
25. Han, T., Kuo, C. H., Sridharan, N., Headings, L. M., Babu, S. S., & Dapino, M. J. (2020). Effect of preheat temperature and post-process treatment on the microstructure and mechanical properties of stainless steel 410 made via ultrasonic additive manufacturing. *Materials Science and Engineering:* A, 769, 138457.
26. Travyanov, A. et al. (2022). Manufacturing of Ti6Al4V alloy part reinforced by silicon carbide fibers by laser powder bed fusion with following hot isostatic pressing. *The International Journal of Advanced Manufacturing Technology*, 119(1), 1331–1341.
27. Seifi, M. et al., (2017). Effects of HIP on microstructural heterogeneity, defect distribution and mechanical properties of additively manufactured EBM Ti-48Al-2Cr-2Nb. *Journal of Alloys and Compounds*, 729, 1118–1135.
28. Yu, H., Li, F., Wang, Z., & Zeng, X. (2019). Fatigue performances of selective laser melted Ti-6Al-4V alloy: Influence of surface finishing, hot isostatic pressing and heat treatments. *International Journal of Fatigue*, 120, 175–183.
29. Cutolo, A. et al. (2018). Influence of layer thickness and post-Process treatments on the fatigue properties of CoCr scaffolds roduced by laser powder bed fusion. *Additive Manufacturing*, 23, 498–504.
30. Wu, M. W., Chen, J. K., Lin, B. H., & Chiang, P. H. (2017). Improved fatigue endurance ratio of additive-manufactured Ti-6Al-4V lattice by hot isostatic pressing. *Materials & Design*, 134, 163–170.
31. Liu, S., & Guo, H. (2020). Influence of hot isostatic pressing (HIP) on mechanical properties of magnesium alloy produced by selective laser melting (SLM). *Materials Letters*, 265, 127463.
32. Schneller, W. et al. (2019). Effect of HIP treatment on microstructure and fatigue strength of selectively laser melted AlSi10Mg. *Journal of Manufacturing and Materials Processing*, 3(1), 16.
33. Hirata, T., Kimura, T., & Nakamoto, T. (2020). Effects of hot isostatic pressing and internal porosity on the performance of selective laser melted AlSi10Mg alloys. *Materials Science and Engineering: A*, 772, 138713.
34. Tran, M. T. et al. (2021). Effect of hot isostatic pressing on the cryogenic mechanical properties of CrCoNi medium entropy alloy processed by direct energy deposition. *Materials Science and Engineering: A*, 828, 142110.

35. Rosenthal, I., Shneck, R., & Stern, A. (2018). Heat treatment effect on the mechanical properties and fracture mechanism in AlSi10Mg fabricated by additive manufacturing selective laser melting process. *Materials Science and Engineering: A*, 729, 310–322.
36. Larrosa, N. O. et al. (2018). Linking microstructure and processing defects to mechanical properties of selectively laser melted AlSi10Mg alloy. *Theoretical and Applied Fracture Mechanics*, 98, 123–133.
37. Jiang, R., Mostafaei, A., Pauza, J., Kantzos, C., & Rollett, A. D. (2019). Varied heat treatments and properties of laser powder bed printed Inconel 718. *Materials Science and Engineering:* A, 755, 170–180.
38. Schneider, J. (2020). Comparison of microstructural response to heat treatment of Inconel 718 prepared by three different metal additive manufacturing processes. *JOM: The journal of the Minerals, Metals & Materials Society*, 72(3), 1085–1091.
39. Tillmann, W. et al. (2015). Functional encapsulation of laser melted Inconel 718 by Arc-PVD and HVOF for post compacting by hot isostatic pressing. *Powder Metallurgy*, 58(4), 259–264.
40. Poulin, J. R., Brailovski, V., & Terriault, P. (2018). Long fatigue crack propagation behavior of Inconel 625 processed by laser powder bed fusion: Influence of build orientation and post-processing conditions. *International Journal of Fatigue*, 116, 634–647.
41. Åsberg, M. et al. (2019). Influence of post treatment on microstructure, porosity and mechanical properties of additive manufactured H13Tool steel. *Materials Science and Engineering: A*, 742, 584–589.
42. Atkinson, H. V., & Davies, S. (2000). Fundamental aspects of hot isostatic pressing: An overview. *Metallurgical and Materials Transactions A*, 31(12), 2981–3000.
43. Ardi, D. T. et al. (2020). Effects of post-processing route on fatigue performance of laser powder bed fusion Inconel 718. *Additive Manufacturing*, 36, 101442.
44. Cheng, W., Yongbin, L., Long, W., & Chuanli, W. (2020). Dislocation-based study on the influences of shot peening on fatigue resistance. *Surface and Coatings Technology*, 383, 125247.
45. Sawicki, J. et al. (2018). Influence of single-stage and duplex shot peening on surface roughness and residual stresses in Al Mg5 Mn1 Sc0,8 Zr0,4 alloy. *Archives of Metallurgy and Materials*, 63(1), 505–511.
46. Qian, M. et al. (2016). Additive manufacturing and postprocessing of Ti-6Al-4V for superior mechanical properties. *MRS Bulletin*, 41(10), 775–784.
47. Uzan, N. E. et al. (2018). On the effect of shot-peening on fatigue resistance of AlSi10Mg specimens fabricated by additive manufacturing using selective laser melting (AM-SLM). *Additive Manufacturing*, 21, 458–464.
48. Damon, J. et al. (2018). Process dependent porosity and the influence of shot peening on porosity morphology regarding selective laser melted AlSi10Mg parts. *Additive Manufacturing*, 20, 77–89.
49. Łach, Ł. (2024). Recent advances in laser surface hardening: Techniques, modeling approaches, and industrial applications. *Crystals*, 14, 726. https://doi.org/10.3390/CRYST14080726.
50. Nemova, G. (2024). Brief review of recent developments in fiber lasers. *Applied Sciences*, 14, 2323. https://doi.org/10.3390/APP14062323.
51. Buling, A., Sändker, H., Stollenwerk, J., Krupp, U., & Hamann-Steinmeier, A. (2016). Laser surface pretreatment of 100Cr6 bearing steel – Hardening effects and white etching zones. *Applied Surface Science*, 378, 564–571. https://doi.org/10.1016/J.APSUSC.2016.03.088.
52. Cholkar, A. et al. (2024). Combining ultrafast laser texturing and laser hardening to enhance surface durability by improving hardness and wear performance. *Advanced Engineering Materials*, 26, 2401183. https://doi.org/10.1002/ADEM.202401183.

53. Roehling, J. D. et al. (2019). Reducing residual stress by selective large-area diode surface heating during laser powder bed fusion additive manufacturing. *Additive Manufacturing*, 28, 228–235. https://doi.org/10.1016/J.ADDMA.2019.05.009.
54. Steen, W. M., & Mazumder, J. (2010). *Laser material processing*. Fourth ed. Springer London. https://doi.org/10.1007/978-1-84996-062-5/COVER.
55. Paschotta, R. Dr. (2008). Continuous-wave operation. *Encyclopedia of Laser Physics and Technology*. Available at: https://www.rp-photonics.com/continuous_wave_operation.html (Accessed on 10/12/2025)
56. Basak, A., & Das, S. (2016). Epitaxy and microstructure evolution in metal additive manufacturing. *Annual Review of Materials Research*, 46, 125–149. https://doi.org/10.1146/ANNUREV-MATSCI-070115-031728/CITE/REFWORKS.
57. Cholkar, A. et al. (2023). Advances in laser-based surface texturing for developing antifouling surfaces: A comprehensive review. *Applied Surface Science Advances*, 18, 100513. https://doi.org/10.1016/J.APSADV.2023.100513.
58. Lesykwang, J., Ding, Y., Hu, X., Qian, L., Wang, Y., & Li F. (2024). Process research of surface laser phase transformation hardening for 42CrMo material. *Photonics*, 11, 1205. https://doi.org/10.3390/PHOTONICS11121205.
59. Lesyk, D., Hruska, M., Mordyuk, B., Kochmanski, P., & Powalka, B. (2023). Robot-assisted 3D laser surface hardening of medium-carbon steel: Surface roughness parameters and hardness. In Karabegovic, I., Kovačević, A., & Mandzuka, S. (Eds.), *New technologies, development and application VI. NT 2023*. Lecture Notes in Networks and Systems, vol 687. Springer, Cham, 45–53. https://doi.org/10.1007/978-3-031-31066-9_5.
60. Leng, Y. et al. (2021). Laser surface hardening of SS316L. *IOP Conference Series: Materials Science and Engineering*, 1070, 012107. https://doi.org/10.1088/1757-899X/1070/1/012107.
61. Babu, P. D., Balasubramanian, K. R., & Buvanashekaran, G. (2011). Laser surface hardening: A review. *International Journal of Surface Science and Engineering*, 5, 131–51. https://doi.org/10.1504/IJSURFSE.2011.041398.
62. Jia, Z.-Y. et al. (2021). Effect of process parameters on the hardness of laser surface textured 5A06 aluminum alloy. *Journal of Materials Engineering and Performance*, 30, 5858–5867. https://doi.org/10.1007/S11665-021-05840-Y/FIGURES/6.
63. Toyserkani, E., Khajepour, A., & Corbin, S. F. (2004). *Laser cladding*. CRC Press. https://doi.org/10.1201/9781420039177.
64. Moradi, M., Karami Moghadam, M., & Shamsborhan, M. (2020). How the laser beam energy distribution effect on laser surface transformation hardening process; Diode and Nd:YAG lasers. *Optik (Stuttg)* 204, 163991. https://doi.org/10.1016/J.IJLEO.2019.163991.
65. Lu, Y., Ehle, L. C., Richter, S., & Radel, T. (2021). Influence of multi-pass laser hardening of normalized AISI 4140 on the grain size. *Surface and Coatings Technology*, 421, 127434. https://doi.org/10.1016/J.SURFCOAT.2021.127434.
66. Mutlu, M., Kacar, E., Akman, E., Akkan, C. K., Demir, P., & Demir, A. (2009). Effects of the laser wavelength on drilling process of ceramic using Nd:YAG laser. *JLMN-Journal of Laser Micro/Nanoengineering*, 4, 84. https://doi.org/10.2961/jlmn.2009.02.0002.
67. Assuncao, E., & Williams, S. (2013). Comparison of continuous wave and pulsed wave laser welding effects. *Optics and Lasers in Engineering*, 51, 674–680. https://doi.org/10.1016/J.OPTLASENG.2013.01.007.
68. Wang, J. et al. (2024). A comprehensive review of metal laser hardening: Mechanism, process, and applications. *The International Journal of Advanced Manufacturing Technology*, 134, 5087–5115. https://doi.org/10.1007/S00170-024-14463-1.

69. Volpp, J., Dewi, H. S., Fischer, A., & Niendorf, T. (2020). Influence of complex geometries on the properties of laser-hardened surfaces. *The International Journal of Advanced Manufacturing Technology*, 107, 4255–4260. https://doi.org/10.1007/S00170-020-05324-8/FIGURES/7.
70. Goebel, G., Bonss, S., Brenner, B., & Beyer, E. (2002). Computational process parameter optimization for laser beam transformation hardening. In *ICALEO 2002 - 21st International Congress on Applications of Laser and Electro-Optics, Congress Proceedings*. https://doi.org/10.2351/1.5065604/893700.
71. Bouland, C., Urlea, V., Beaubier, K., Samoilenko, M., & Brailovski, V. (2019). Abrasive flow machining of laser powder bed-fused parts: Numerical modeling and experimental validation. *Journal of Materials Processing Technology*, 273(June), 116262. https://doi.org/10.1016/j.jmatprotec.2019.116262.
72. Venettacci, S., Ponticelli, G. S., & Guarino, S. (2021). Fluidised bed finishing process for aeronautical applications: Environmental and technical-economic assessment. *Journal of Cleaner Production,* 299(May), 126900. https://doi.org/10.1016/j.jclepro.2021.126900.
73. Yates, J. G. (1993). Fluidization engineering: Second edition. *Powder Technology*, 74(2), 197. https://doi.org/10.1016/0032–5910(93)87011-C.
74. Vogt, E. T. C., & Weckhuysen, B. M. (2015). Fluid catalytic cracking: Recent developments on the grand old lady of zeolite catalysis. *Chemical Society Reviews*, 44 (20), 7342–7370. https://doi.org/10.1039/c5cs00376h.
75. Cai, P., Chen, L., Van Egmond, J., & Tilston, M. (2010). Some recent advances in fluidized-bed polymerization technology. *Particuology*, 8(6), 578–581. https://doi.org/10.1016/j.partic.2010.09.006.
76. Jiang, K., Du, X., Zhang, Q., Kong, Y., Xu, C., & Ju, X. (2021). Review on gas-solid fluidized bed particle solar receivers applied in concentrated solar applications: Materials, configurations and methodologies. *Renewable and Sustainable Energy Reviews*, 150 (July), 111479. https://doi.org/10.1016/j.rser.2021.111479.
77. Wu, C., Yang, H., He, X., Hu, C., Yang, L., & Li, H. (2022). Principle, development, application design and prospect of fluidized bed heat exchange technology: comprehensive review. *Renewable and Sustainable Energy Reviews*, 157, 112023. https://doi.org/10.1016/j.rser.2021.112023.
78. Kaminsky, W. (2021). Chemical recycling of plastics by fluidized bed pyrolysis. *Fuel Communications*, 8(July), 100023. https://doi.org/10.1016/j.jfueco.2021.100023.
79. Xue, Y., Zhou, S., Brown, R. C., Kelkar, A., & Bai, X. (2015). Fast pyrolysis of biomass and waste plastic in a fluidized bed reactor. *Fuel*, 156, 40–46. https://doi.org/10.1016/j.fuel.2015.04.033.
80. Raganati, F., Scherillo, F., Squillace, A., Chirone, R., & Ammendola, P. (2018). Improvement of the manufacturing process of tungsten carbide-cobalt hard metals by the application of sound assisted fluidization for the mixing of the powders. *Industrial and Engineering Chemistry Research*, 57(1), 414–424. https://doi.org/10.1021/acs.iecr.7b04627.
81. El Hassanin, A, Scherillo, F., Prisco, U., Sansone, R., & Astarita, A. (2020). Selective laser melting of Cu-Inconel 718 powder mixtures. *Journal of Manufacturing Processes*, 59(July), 679–689. https://doi.org/10.1016/j.jmapro.2020.10.039.
82. El Hassanin, A., Teresa Silvestri, A., Napolitano, F., Borrelli, D., Caraviello, A., & Astarita, A. (2023). Investigation of the laser-related parameters in laser-powder bed fusion of Inconel 718-Cu blends at fixed volumetric energy density. *Journal of Manufacturing Processes*, 99(July), 456–468. https://doi.org/10.1016/j.jmapro.2023.05.068.

83. El Hassanin, A. et al. (2021). Laser-powder bed fusion of pre-mixed Inconel718-Cu powders: An experimental study. *Journal of Manufacturing Processes*, 71(September), 329–344. https://doi.org/10.1016/j.jmapro.2021.09.028.
84. Jiang, S. et al. (2024). Advancements of the Fluidized Bed Fenton (FBF) technology for wastewater treatment: Mechanism, mass and heat transfer. *Journal of Environmental Management*, 362(June), 121325. https://doi.org/10.1016/j.jenvman.2024.121325.
85. Sivakumar, R., Saravanan, R., Elaya Perumal, A., & Iniyan, S. (2016). Fluidized bed drying of some Agro products - A review. *Renewable and Sustainable Energy Reviews*, 61, 280–301. https://doi.org/10.1016/j.rser.2016.04.014.
86. Kjaer Jepsen, P., De Juan, L. M., Boje, A., Sarangi, S., von Corswant, C., & Sasic, S. (2024). Wurster fluidised-bed coating: Coarse-graining technique within CFD-DEM in conjunction with heat and mass transfer. *Powder Technology*, 443(March), 119901. https://doi.org/10.1016/j.powtec.2024.119901.
87. Dewettinck, K., & Huyghebaert, A. (1999). Fluidized bed coating in food Technology. *Food Science and Technology*, 10, 163–168.
88. Song, Y., Zhou, T., Bai, R., Zhang, M., & Yang, H. (2024). Assessment of the coating quality in a top-spray fluidized bed coater: An experimental study. *Powder Technology*, 439(March), 119663. https://doi.org/10.1016/j.powtec.2024.119663.
89. Barletta, M. (2006). A new technology in surface finishing: Fluidized bed machining (FBM) of aluminium alloys. *Journal of Materials Processing Technology*, 173(2), 157–165. https://doi.org/10.1016/j.jmatprotec.2005.11.020.
90. Barletta, M., Guarino, S., Rubino, G., & Tagliaferri, V. (2007). Progress in fluidized bed assisted abrasive jet machining (FB-AJM): Internal polishing of aluminium tubes. *International Journal of Machine Tools and Manufacture*, 47 (3–4), 483–495. https://doi.org/10.1016/j.ijmachtools.2006.06.005.
91. Barletta, M. (2009). Progress in abrasive fluidized bed machining. *Journal of Materials Processing Technology*, 209(20), 6087–6102. https://doi.org/10.1016/j.jmatprotec.2009.04.009.
92. Finne, I. (1960). Erosion of surfaces. *Wear*, 3, 87–103. https://doi.org/10.1016/0043–1648(60)90055-7.
93. Khrushchev, M. M., Babichev, M. A., National Science Foundation (U.S.), & U S Atomic Energy Commission. (1953). *Resistance to abrasive wear and the hardness of metals*. NSF-Tr. United States Atomic Energy Commission, Technical Information Service. https://books.google.it/books?id=7QJZbPuokuoC.
94. Zum Gahr, K. H. (1998). Wear by hard particles. *Tribology International*, 31(10), 587–596. https://doi.org/10.1016/S0301–679X(98)00079-6.
95. Sanaei, N., & Fatemi, A. (2020). Analysis of the effect of surface roughness on fatigue performance of powder bed fusion additive manufactured metals. *Theoretical and Applied Fracture Mechanics*, 108(April), 102638. https://doi.org/10.1016/j.tafmec.2020.102638.
96. Atzeni, E., Barletta, M., Calignano, F., Iuliano, L., Rubino, G., & Tagliaferri, V. (2016). Abrasive fluidized bed (AFB) finishing of AlSi10Mg substrates manufactured by direct metal laser sintering (DMLS). *Additive Manufacturing*, 10, 15–23. https://doi.org/10.1016/j.addma.2016.01.005.
97. Atzeni, E., Rubino, G., Salmi, A., & Trovalusci, F. (2020). Abrasive fluidized bed finishing to improve the fatigue behaviour of Ti6Al4V parts fabricated by electron beam melting. *The International Journal of Advanced Manufacturing Technology*, 110(1–2), 557–567. https://doi.org/10.1007/s00170-020-05814-9.
98. Barletta, M., Tagliaferri, V., Trovalusci, F., Veniali, F., & Gisario, A. (2013). The mechanisms of material removal in the fluidized bed machining of polyvinyl chloride substrates. *Journal of Manufacturing Science and Engineering*, 135(1), 1–11. https://doi.org/10.1115/1.4007956.

99. Troiano, M. et al. (2024). Fluidized bed finishing of additively manufactured objects: The influence of operating parameters. *Powder Technology*, 432, 119115. https://doi.org/10.1016/j.powtec.2023.119115.
100. Atydu, T. (1954). Experiments on a gravity-free dispersion of large solid spheres in a Newtonian fluid under shear. *Proceedings of the Royal Society of London. Series A. Mathematical and Physical Sciences*, 225(1160), 49–63. https://doi.org/10.1098/rspa.1954.0186.
101. Sunay, N., Kaya, M., & Kaynak, Y. (2020). Properties of parts fabricated by additive manufacturing. *Sigma Journal of Engineering and Natural Sciences*, 38(4), 2027–2042.
102. Łyczkowska, E., Szymczyk, P., Dybała, B., & Chlebus, E. (2014). Chemical polishing of scaffolds made of Ti–6Al–7Nb alloy by additive manufacturing. *Archives of Civil and Mechanical Engineering*, 14 (4), 586–594. https://doi.org/10.1016/j.acme.2014.03.001.
103. Boban, J., Ahmed, A., Jithinraj, E. K., Azizur Rahman, M., & Rahman, M. (2022). Polishing of additive manufactured metallic components: Retrospect on existing methods and future prospects. *The International Journal of Advanced Manufacturing Technology*, 121(1–2), 83–125. https://doi.org/10.1007/s00170-022-09382-y.
104. Tyagi, P., Goulet, T., Riso, C., & Garcia-Moreno, F. (2019). Reducing surface roughness by chemical polishing of additively manufactured 3D printed 316 stainless steel components. *The International Journal of Advanced Manufacturing Technology* 100(9–12), 2895–2900. https://doi.org/10.1007/s00170-018-2890-0.
105. Scherillo, F., Manco, E., El Hassanin, A., Franchitti, S., Pirozzi, C., & Borrelli, R. (2020). Chemical surface finishing of electron beam melted Ti6Al4V using HF-HNO3 solutions. *Journal of Manufacturing Processes*, 60(December), 400–409. https://doi.org/10.1016/j.jmapro.2020.10.033.
106. Sutter, E. M. M., & Goetz-Grandmont, G. J. (1990). The behaviour of titanium in nitric-hydrofluoric acid solutions. *Corrosion Science*, 30(4–5), 461–476. https://doi.org/10.1016/0010–938X(90)90051-6.
107. Sutter, E. M. M., Hlawka, F., & Cornet, A. (1990). Comparative behavior of titanium and zirconium in hydrofluoric-nitric acid pickling solutions. *CORROSION*, 46(7), 537–544. https://doi.org/10.5006/1.3585144.
108. Kim, D. E., & Suh, N. P. (1992). Chemical smoothing of rough metal surfaces. *Journal of Engineering for Industry*, 114(4), 421–426. https://doi.org/10.1115/1.2900693.
109. Scherillo, F. (2019). Chemical surface finishing of AlSi10Mg components made by additive manufacturing. *Manufacturing Letters*, 19, 5–9. https://doi.org/10.1016/j.mfglet.2018.12.002.

9 Bimetallic Material Thin-Walled Structures: Manufacturing and Applications

Sarika Kumari

9.1 INTRODUCTION

Advanced manufacturing has recently observed increasing interest in bimetallic materials, which are hybrid materials that consist of two different metals or their alloys [1]. These materials give higher performance, such as mechanical strength, corrosion resistance, and thermal stability, and they have lower cost than traditional materials, so these materials are precious and useful for industries that need multifunctional parts [2]. Lightweight structures, often known as thin-walled structures, are used in aerospace, automobile, marine, and biomedical fields. In such applications, bimetallic thin-walled components offer features such as lightweight, high strength-to-weight ratio, low fuel consumption, and longer life [2–3]. Due to the possibilities offered by the techniques of advanced manufacturing, the fabrication of bimetallic thin-walled structures with highly specific geometric designs has been made possible through precise control of the material layout and the possibility of varying mechanical properties of the material at the micro level. With this advancement, it is now possible to create a large number of components made out of more than one material, which were previously hard to manufacture.

9.2 FUNDAMENTALS OF BIMETALLIC MATERIALS

Bimetallic materials are multilayer structures that consist of two dissimilar metals or their alloys, at least in terms of their physical, chemical, or mechanical properties. These materials are purposely developed to increase the total performance of these parts through the use of distinct metals. There are two primary types of bimetallic structures: homogeneous and heterogeneous, as shown in Figure 9.1 [3–4]. In based bimetallic structures, the interface of the two metals is smoothly

 DOI: 10.1201/9781003571209-9

Characteristic	Homogeneous	Heterogeneous
Interface	Smoothly formed	Separated
Material Joining	Diffusion bonding or additive manufacturing	Cladding or welding
Material Pairs	Steel-copper, aluminium-titanium	Nickel-based alloys-stainless steel
Mechanical Properties	Tensile strength, fatigue resistance	Wear rate
Other Characteristics	Corrosion resistance	Thermal stability
Cost Efficiency	Expensive metal	Performance enhancement

FIGURE 9.1 Comparison between homogeneous and heterogeneous bimetallic structures

formed through diffusion bonding or additive manufacturing (AM) techniques. On the other hand, heterogeneous bimetallic structures have an interface that separates the two metals in which they are laid or bonded through procedures like cladding or welding [5–6].

There are often used material pairs, for example, steel-copper, aluminum-titanium, and nickel-based alloys-stainless steel [7]. Why each combination is best explained by the fact that each combination provides certain benefits depending on its placement in use. For example, free ends are those whose items are made from steel and copper, the latter being preferred for its increased conductivity; hence, steel is used to enhance the mechanical strength. There has always been a high demand for aluminum-titanium bimetallic materials in aerospace and automobile industries due to low density, high tensile strength, and corrosion resistance. The characteristics of such multicomponent materials are presented in the ability to use bimetallic materials in diverse industry fields. Superior mechanical properties, including tensile strength, fatigue resistance, and wear rate of the biomaterial, are obtained by joining two metals with the respective properties [8–9]. Other important characteristics, including corrosion resistance and thermal stability, are also optimized, as one metal serves as a shield against severe environmental conditions while the other can retain strength at high temperatures [10–11]. Furthermore, bimetallic material is cost-efficient in using expensive metal, which is a bonus for production expenses and at the same time enhances the performance of the part [12]. Such properties make bimetallic thin-walled structures a valuable and prospective solution for industries with demands for lightweight and multifunctional components with high strength.

9.3 ADDITIVE MANUFACTURING TECHNIQUES FOR BIMETALLIC THIN-WALLED STRUCTURES

AM has opened new opportunities to create thin-walled three-state bimetallic structures using different metals, which would have been impossible to create earlier. There have been several approaches to creating the necessary properties and functionalities of bimetallic components in AS and AM technologies. This section looks at AM techniques, including laser-based methods, wire arc additive manufacturing (WAAM), powder bed fusion (PBF), and manufacturing techniques, comparing their processes, benefits, and drawbacks as shown in Table 9.1.

TABLE 9.1
Additive manufacturing techniques used in bimetallic material printing

AM Techniques	Description	Key features	Limitations	Applications
Laser-Based AM (Laser Cladding, DED, SLM)	Uses a laser to fuse material layer by layer; includes Laser Cladding, Direct Energy Deposition (DED), and Selective Laser Melting (SLM)	• High precision & flexibility • Strong metallurgical bonds • Suitable for intricate designs • Good interfacial adhesion	• Thermal stresses due to CTE mismatch • Cracking & poor adhesion if not controlled • Material transport must be stable	• Repair of components • Complex bimetallic parts • Functional layer deposition
Wire Arc Additive Manufacturing (WAAM)	Uses an electric arc to melt metal wire, depositing it layer by layer	• High deposition rates • Cost effective for large parts • Good material utilization • Suitable for dissimilar materials	• Thermal dissipation issues • Interface bonding challenges • Requires post-treatment • Defects like porosity and poor fusion	• Aerospace • Automotive • Marine structures
Powder Bed Fusion (PBF)	Fuses metal powder using a laser/electron beam in a powder bed	• High precision and design freedom • Suitable intricate bimetallic feature	• Material compatibility concerns • Varying melting points and CTE • Powder mixing and thermal control required	• Complex parts with fine details • Precision bimetallic structures

9.3.1 Laser-Based Additive Manufacturing (Laser Cladding, DED, SLM)

The collinear laser AM processes that include laser cladding, direct energy deposition (DED), and selective laser melting (SLM) are appropriate for building bimetallic thin-walled structures because they are precise and flexible techniques for creating a manufacturing part [13–14]. In these processes, a laser beam fuses and accumulates the material layer by layer to form the reinforced structure. Laser cladding can be described as a process that deposits layers of a protective or functional material on a base material, making it ideal for repairing or building on existing parts. In DED, two or more metal powders or wires can be directly deposited into a base material or another layer, in which there is a metallurgical bond between the material being deposited and another material [15]. SLM is most suitable in applications that require accurate positioning of the two metals in shapes of intricate designs [16].

Laser-based methods also present themselves as having good advantages, such as high accuracy, good interfacial adhesion, and better geometries. But they are not without problems, as we observe in the examples below: thermal expansion coefficient variation in metals produces stresses and some cracks. Furthermore, it demonstrates that steady transportation of the material and achieving appropriate adhesion at the joint are important for the fabrication of flawless bimetallic products.

9.3.2 Wire Arc Additive Manufacturing

The WAAM technique has been adopted by researchers, as the fabrication of large-scale bimetallic thin-walled structures is more economical [17–18]. In WAAM, an electric arc is used to melt a metal wire, and the layers of the component being constructed are built up using this wire. This process is most preferable for structural members for applications in aerospace [19], automobile [20], and marine industries [21], where large and strong components with excellent corrosion resistance are desirable. Wire feed rate, arc voltage, current, and travel speed are the major parameters that define the WAAM process. These parameters have to be well regulated to guarantee the correct interaction within dissimilar metal joints and to prevent such defects as porosity and inadequate fusion. The uses of WAAM include high deposition rates, which give high material utilization rates, low wastage, and the use of dissimilar materials. However, issues like thermal dissipation, interface connection, and post-treatment processes remain an issue hindering the quality of bimetallic structures for use [22].

9.3.3 Powder Bed Fusion

PBF is one of the most commonly used AM processes for manufacturing precise bimetallic thin-walled structures. In this process, a thin layer of metal powder is leveled on a build platform, and a laser or electron beam simply fuses the material to make the object's shape. This continues on a successive layering until a building is firm [23–25]. Of the PBF techniques, SLM is highly suitable for creating intricate bimetallic parts with small details. Even though PBF seems to bring excellent precision and design freedom, material compatibility becomes a critical issue, especially when dealing with bimetallic parts [26]. Principal types of these issues include the dissimilar melting points and thermal conductivities of the two metals, as well as differing coefficients of thermal expansion. Admixing the performance of the powder and controlling the thermal conditions during the process contribute to better development of bimetallic thin-walled structures.

9.3.4 Hybrid Manufacturing Techniques

The integration of AM with subtractive processes like milling and turning creates thin-walled bimetallic components with high precision and surface finish [27]. Also, in hybrid manufacturing, functional features are produced in layers by additive processes, and strict geometries are executed by subtraction processes such as CNC milling at high tolerances and surface finishing [28–30]. The one benefit that is usually associated with hybrid manufacturing is that it provides the benefits of both manufacturing processes, offering higher accuracy, smoother surfaces, and improved joints. This makes it an ideal tool for the repair and reconditioning of parts, in addition to allowing the modification of existing parts, and is therefore suitable for industrial use [31]. However, the simultaneous performance of both additive and subtractive processes can be a challenge due to the requirement of various tools and globally consistent integration, and thus, it can be more time-consuming and expensive than individual AM technologies.

9.4 CHALLENGES IN MANUFACTURING BIMETALLIC THIN-WALLED STRUCTURES

9.4.1 Material Compatibility Issues

The key issue when assembling bi-metallic (BM) thin-walled structures is the compatibility of the materials, including their thermal and chemical characteristics.

9.4.1.1 Thermal Expansion Mismatch

If one forms an alloy of two different metals, they will realize that they have different coefficients of thermal expansion. When built, the materials are heated, and differentials in the build process make the material expand and contract at different rates. Such a scenario harbors micro-stresses, formation of cracks or delamination between the interface, and affects the structural characteristics of the given component. For instance, it is not advisable to integrate aluminum with steel because of the large disparity in their thermal coefficient of expansion [32].

9.4.1.2 Diffusion Bonding Challenges

In structures made from both metals, the connection must be stronger at the interface of the two metals. In diffusion bonding, atoms of both metals rearrange themselves at the interface, thus producing a metallurgical joint. However, this process can be hindered by differences in melting points, chemical incompatibilities, or the presence of an intermetallic phase of lower melting point, which is deleterious to the ultimate bond strength. This is true because it is very important to produce a robust and defect-free interface for the bimetallic structures [33–34].

9.4.2 Process-Related Challenges

The technical difficulties inherent in the AM process become apparent at the level of bimetallic thin-walled structures and have to be addressed systematically to achieve high-quality results.

9.4.2.1 Heat Control and Residual Stress Management

Steady-state thermal loading occurs in processes that use, for example, laser or electron beams to deposit layers of material. This results in temperature differences within the structure, which subsequently give rise to stresses that can cause warping and/or cracking of the component or else distortion of the structure. The application of heat and cooling rates is critical, as they can affect the final level of stress, as well as the stability of shape in the constructed bimetallic structure. Some of the methods include heating the build platform and applying controlled cooling to reduce these stresses [35].

9.4.2.2 Layer Adhesion and Interfacial Bonding

Sealing one metal to another with great interaction between the two conjoining materials is also a success factor in engineering thin-walled structures of bimetallic construction. Poor bonding may lead to the formation of poor interfaces, whereupon loading the structure is likely to delaminate or fail. It is possible to influence the characteristics of material deposition and bonding through control of such parameters as the material feed rate, laser power, and scanning velocity [36–37].

9.4.3 Post-Processing Challenges

9.4.3.1 Machining and Surface Finishing of Bimetallic Structures

Machining, grinding, or polishing may be carried out during post-processing in order to attain the right surface finish and dimension. Such an arrangement is challenging when post-processing, particularly when the bimetallic structure uses two dissimilar metals with different hardness and thermal conductivity. For instance, when machining a structure of copper and steel, different cutting speeds of the tool and materials are used for each metal, raising the likelihood of tool wear and time consumption. It also requires that the interface between the two metals be free from defects and that there be a low voltage gradient across the point of transition between the two materials, because if there is a high voltage gradient, it causes stress concentration, which in turn leads to corrosion [38–40].

9.5 CASE STUDIES OF BIMETALLIC THIN-WALLED STRUCTURES

9.5.1 Aerospace Applications

Recently, the thermal buckling behavior of thin-walled bimetallic box beams and frame structures was studied using a published *Materials* paper. In this research, stability analysis under uniform temperature increases of these structures was conducted based on the Euler–Bernoulli–Vlasov beam theory, which incorporates large rotations and small strains. It was found that the critical buckling temperature and post-buckling responses are dominated by boundary parameters, beam length, and material thickness ratios. This is important for the aerospace industry, which often faces components in different thermal fields [41]. Advancements in AM mean that bimetallic structures with complex geometries can now be manufactured. *Virtual and Physical Prototyping* has reviewed various AM processing strategies, characterization methods, and bimetallic structures applications. The review noted the advantages of AM for the realization of components meeting the stringent aerospace applications requirements for weight reduction and performance improvements [42]. The solution proposed for designing and optimizing complex large-scale thin-walled structures was also published in the research paper in Materials. In the aerospace applications, the study focused on the main support structures, with structural integrity and weight efficiency being the main points for the balance. The application of bimetallic thin-walled structures in the aerospace sector is advancing with the help of the proposed design and optimization strategies [43].

9.5.2 Automotive Applications

In recent years, the automotive industry has been looking toward bimetallic materials in exhaust systems and engine components, as their ability to integrate the good properties of different metals has been exploited. These materials give the structure improved mechanical strength, thermal stability, and corrosion resistance while maintaining a lightweight structure that is essential for minimizing fuel consumption and reducing emissions. Bimetallic components are used in the exhaust systems to compensate for thermal misfit between different metal parts. For instance, bimetallic washers are placed between the exhaust manifold and studs so that the exhaust manifold does what it is supposed to do and in a manner that ensures long-term durability, yet will negate any difference in thermal expansion due to constant use. To reduce the weight of each system, metals such as stainless steel and aluminum are combined in an exhaust system to maintain high temperature resistance and mechanical strength. Some research has been carried out integrating bimetallic materials into exhaust components to improve thermal efficiency and decrease heat-cycling-induced material degradation [44]. Bimetallic structures also greatly benefit engine components. For example, engine cradles made of bimetallic steel tubes instead of aluminum extrusions are integrated, which results in a lightweight yet highly rigid structure. By combining the advantage of weight in better fuel economy and still improving load-bearing capacity, this hybrid approach is utilized. Bimetallic bearings and cylinder liners may be used in high-performance applications to combine surfaces that are resistant to wear with materials that have better heat dissipation properties in order to extend the lifespan and efficiency of the engine [45]. While the advantages related to dissimilar metal use in automotive applications are numerous, use of dissimilar metals in automotive applications is challenged mainly by conventional bimetallic corrosion, where one metal can reduce and thus react electrochemically with electromotive force and mobilizes electrons from the other metal ion. Such a situation may result in premature material degradation and lower component lifetime.

Challenges such as these are overcome by proper selection of materials, protective coatings, and advanced joining techniques (e.g., diffusion bonding and laser welding). These processes are continuing to be researched to optimize the development of these processes for enhanced performance and reliability of bimetallic automotive components [46].

9.5.3 Marine and Oil and Gas Applications

The marine and oil and gas sectors demand corrosion-resistant piping materials because their operational conditions include unforgiving environments. Cupronickel alloys serve marine engineering purposes because they resist seawater damage and macrofouling attacks effectively. Ship hulls, seawater cooling systems, and desalination plant tubing exclusively use these alloys for their applications. The oil and gas sector employs corrosion-resistant alloys, including duplex stainless-steel materials, for its pipeline and fitting needs. Offshore drilling, as well as subsea pipelines, benefit from using duplex stainless steels because these materials demonstrate high resistance to chloride-induced stress corrosion cracking, alongside excellent strength performance under deep pressure conditions. Thermoplastic materials combined with multilayer piping systems have expanded usage in marine environments because they offer excellent corrosion prevention and weigh less and last longer than traditional options. The materials must pass several tests for both safety and performance quality to demonstrate their reliable operation in harsh maritime conditions [47–49].

9.5.4 Biomedical Applications

The combination of different materials through multi-material 3D printing in prosthetics enables improved comfort features and functional benefits to prosthetic devices. The combination of flexible and rigid materials makes possible connection areas between prosthetics and body surfaces to become more comfortable, yet structural elements maintain their needed strength and structural integrity. Hospital-developed prosthetics through this methodology result in better patient-specific designs that improve the entire experience for users. Implants achieve multifaceted advantages through 3D printing methods that produce devices made of materials with targeted property changes across their composition (FGMs). The integration between an implant and biological tissue becomes more effective because functionally graded materials (FGMs) incorporate smooth transitions throughout single units. By using FGMs in temporomandibular joint prostheses, medical practitioners achieve improved mandibular movement along with decreased joint forces that deliver superior treatment results to patients [50]. Technology lets producers create implants that exactly match individual patient anatomies. The tailored design reduces surgical risk factors while extending the lifespan, along with better performance of implant systems. The printing process with multiple materials enables the creation of bone in-growth-promoting porous structure elements along with solid structural sections, which deliver vital mechanical strength [51].

9.6 APPLICATIONS OF BIMETALLIC THIN-WALLED STRUCTURES

9.6.1 Structural Applications

Bimetallic thin-walled structures (Figure 9.2) are used in almost all branches of industry where lightweight structures with high mechanical properties are required. Notably, such industries as aerospace, automotive, and construction receive significant advantages from these structures because the weight reduction does not decrease their strength. For instance, in the aerospace industry, aluminum-titanium bimetallic members are adopted since they offer the best mechanical properties at densities, which increases fuel efficiency by lowering the mean weight of the flying structure. In the same way, steel-aluminum multilayer composites for automobile parts, including chassis parts, frames, and reinforcements, improve the car's crash performance while decreasing energy utilization. These structures can also be used in some other structures, like in Bridges, Towers, etc., where using materials like stainless steel and carbon steel together provides strength and also protection from corrosion [52–55].

9.6.2 Thermal Applications

There are good prospects for using thin-walled structures made of bimetals in thermal applications (Figure 9.2), especially in heat exchanging and cooling installations. The dissimilar materials provide a means of heat dissipation, as it is a critical property depending on the means of thermal control within an application. For instance, in heat exchangers, both copper and aluminum bimetallic parts are employed due to the ability of copper to conduct heat and the light weight of aluminum. This combination is especially suitable for automotive radiators, HVAC applications, and power plant cooling applications, among others. Further, bimetallic thin-walled structures are applied to thermocouples and temperature sensors, in the junction of which the formation of electric voltage proportional to temperature difference is observed. They are also used in energy-conserving building systems like composite panels that control indoor temperatures [56–57].

9.6.3 Wear-Resistant Applications

Thin-walled structures of bimetallic products provide wear resistance and are used in cutting tools, molds, and dies for the manufacturing processes, as depicted in Figure 9.2. In these applications, one material imparts hardness and wear resistance, while the other imparts toughness and impact strength. For instance, steel and carbide bimetallic tools; metallurgical, mining, automobile industries use steel and carbide bimetallic tools to handle abrasion, impact, and high temperature while cutting or machining [58]. The same is true of bimetallic molds – models made of proving steel based on wear-resistant coatings are used in injection molding of plastics and have high durability and the possibility of precise reproduction. In industries where the component is in contact with aggressive environments and abrasive substances, bimetallic structures assist in serving a longer lifespan than when the parts are individually assembled, hence cutting maintenance costs and time [59].

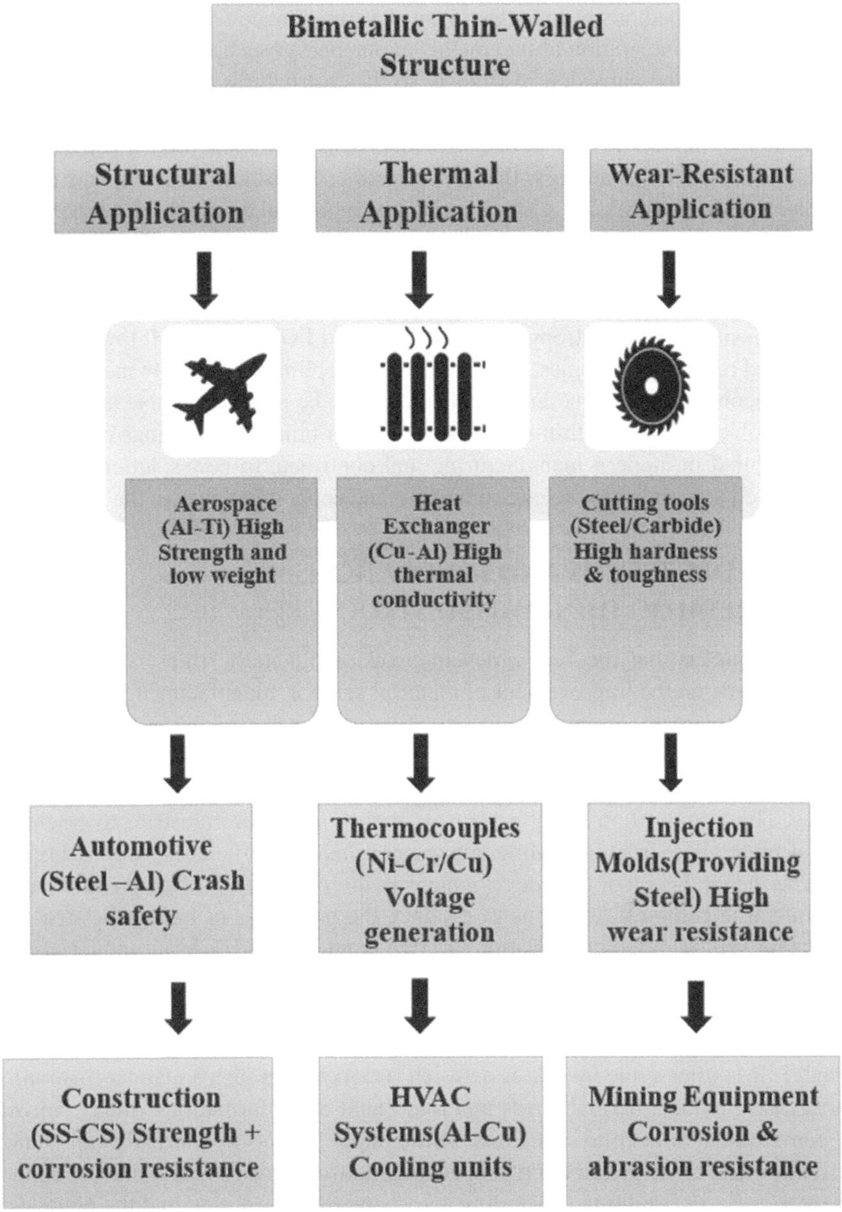

FIGURE 9.2 Applications of bimetallic thin-walled structures

9.7 CONCLUSION

In conclusion, the thin-walled structures of bimetallic materials are considered as one of the good solutions to the industrial needs of multitalented, lightweight, and high-resistance parts. In this chapter, an elaborate discussion of bimetallic materials, properties, and methods of manufacturing structures made from these materials has been given. The integration of two metals in one part gives better strength, improved heat treatment, and corrosion resistance, so these structures can be useful in structural heat and wear areas in astronautic, automotive, and marine industries. However, fabrication of bimetallic thin-walled structures does pose some problems that include compatibility of the materials, the concentration of residual stresses, and possible problems in post-processing – all of which make optimal control of the process. The basic techniques of AM, as well as the incorporation of AI and ML in technology, assist in surmounting these hurdles and offer precise finishing in bimetallic structures. Thus, the further perspective of bimetallic manufacturing will be driven by new material compositions, high-entropy alloys (HEAs), and FGMs, and shift toward sustainability and recycling strategies. With industries adopting eco-friendly measures and a circular economy, such structures shall continue to be more affordable and economically feasible. In general, thin-walled structures of bimetallic coatings can be effectively applied in modern manufacturing and contribute to new solutions in hi-tech uses as well as respond to the environmental and economic issues in the future.

9.8 FUTURE TRENDS AND INNOVATIONS IN BIMETALLIC THIN-WALLED STRUCTURES

It can be stated that the future development of bimetallic thin-walled constructions depends on the improvement of material science, manufacturing technologies, and environmentally friendly practices. New tendencies relate to the creation of new material pairings, AI technologies' application in production, and sustainable approaches to increase performance, decrease expenses, and solve environmental issues. These innovations will foster new opportunities for industries to extend the use of bimetallic structures while achieving reliability, efficiency, and sustainability. One general direction of development is the preparation of new combinations of bimetallic columns. Solutions to enhance the properties of bimetallic structures are being sought in the combination of HEAs and FGMs. HEAs are materials synthesized from multiple elements with nearly balanced atomic fractions; they possess enhanced strength, high thermal stability, and higher combined corrosion resistance, which makes them excellent in high-performance applications [60–62]. On the other hand, FGMs offer a fine interlayer between two metals in such a manner that will, in most cases, reduce thermal stress and interfacial crack formations when compared to conventional bimetallic structures. These improvements in material combinations will allow the construction of optimized and multi-application parts that are suitable to work in harsh conditions, including aerospace, nuclear power, and the biomedical industries. Another major trend is the implementation of modern AM technologies to increase the production of bimetallic thin-walled structures. Artificial Intelligence (AI) and machine learning (ML) are expected to dramatically change the AM process by identifying optimal values of the critical parameters, such as laser power, material flow, and cooling time. AI also can provide much more accurate and real-time estimates of whether a product is likely to have a defect and minimize defects in production lines. Additionally, the development of multi-material processes will allow for direct manufacturing of functional components that integrate bimetallic systems with easy and better control and without time-consuming tool exchanges or add-on processes. Such improvements will greatly improve the speed and precision of multi-metal AM processes [63].

However, novel sustainable concepts and cycles in the application of bimetallic construction are already becoming part of the future application progress in materials and processes. Business sectors are beginning to practice environmental conservation measures to reduce wastage and environmental effects of manufacturing activities. Efficient AM technologies as a means of energy savings, and use of fresher powders, along with the policy of closed cycles, are being adopted to account for the material usage. Furthermore, advances in metal separation technology are being made to enable the recovery of bimetallic parts that have so far been difficult due to the adherence of two distinct metals. High efficiency for the separation process through chemical and electron processes will come in handy in the recovery of valuable metals in a cyclical manner.

REFERENCES

1. Huang, Samuel H., Peng Liu, Abhiram Mokasdar, and Liang Hou. "Additive manufacturing and its societal impact: A literature review." *The International Journal of Advanced Manufacturing Technology* 67 (2013): 1191–1203.
2. Salunkhe, Sachin, and D. Rajamani. "Current trends of metal additive manufacturing in the defense, automobile, and aerospace industries.", In: Sachin Salunkhe, Sergio T. Amancio-filho and J. Paulo Davim (eds) *Advances in Metal Additive Manufacturing*, pp. 147–160. Woodhead Publishing, 2023.
3. Wang, Dingsheng, and Yadong Li. "Bimetallic nanocrystals: Liquid-phase synthesis and catalytic applications." *Advanced Materials* 23, no. 9 (2011): 1044–1060.
4. De, Sudipta, Jiaguang Zhang, Rafael Luque, and Ning Yan. "Ni-based bimetallic heterogeneous catalysts for energy and environmental applications." *Energy & Environmental Science* 9, no. 11 (2016): 3314–3347.
5. Marakatti, Vijaykumar S., and Sebastian C. Peter. "Synthetically tuned electronic and geometrical properties of intermetallic compounds as effective heterogeneous catalysts." *Progress in Solid State Chemistry* 52 (2018): 1–30.
6. Zhou, Ziming, and Yi Ding. "A study on the fatigue performance and corrosion resistance of 304/45 bimetallic composite bolts." *Materials* 16, no. 12 (2023): 4454.
7. Guo, Yanyang, Xiaowei Wu, Guoxiong Ren, Zepeng Liu, Ruidi Yuan, Xiao Yang, and Peng Dong. "Microstructure and properties of copper-steel bimetallic sheets prepared by friction stir additive manufacturing." *Journal of Manufacturing Processes* 82 (2022): 689–699.
8. Chen, Jie, Guijun Bi, Dake Zhao, Di Wang, Mingkang Zhang, and Yunmian Xiao. "Subsequent enhancement of mechanical properties using heat treatment of steel/copper bimetallic structures manufactured by laser powder bed fusion." *Journal of Materials Research and Technology* 33 (2024): 8944–8957.
9. Gadakh, Vijay S., Vishvesh J. Badheka, and Amrut S. Mulay. "Solid-state joining of aluminum to titanium: A review." *Proceedings of the Institution of Mechanical Engineers, Part L: Journal of Materials: Design and Applications* 235, no. 8 (2021): 1757–1799.
10. Raji, Sadiq Abiola, Abimbola Patricia Idowu Popoola, Sisa Leslie Pityana, Olawale Muhmmed Popoola, Fatai Olufemi Aramide, Monnamme Tlotleng, and Nana Kwamina Kum Arthur. "Laser based additive manufacturing technology for fabrication of titanium aluminide-based composites in aerospace component applications." In: Mofid Gorji-Bandpy and Aly-Mousaad Aly (eds) *Aerodynamics*. IntechOpen, 2019.
11. Zhao, Pingyu, Huiyong Ban, Yi-Fei Hu, Kwok-Fai Chung, and Yongjiu Shi. "Residual stress within stainless-clad bimetallic steel welded box sections." *Thin-Walled Structures* 177 (2022): 109395.

12. Hamza, Ali, Kah Chun Ho, and Mieow Kee Chan. "Recent development of substrates for immobilization of bimetallic nanoparticles for wastewater treatment: A review." *Environmental Science and Pollution Research* (2024), 31, 40873–40902.
13. Paydas, Hakan, Anne Mertens, Raoul Carrus, Jacqueline Lecomte-Beckers, and J. Tchoufang Tchuindjang. "Laser cladding as repair technology for Ti–6Al–4V alloy: Influence of building strategy on microstructure and hardness." *Materials & Design* 85 (2015): 497–510.
14. Zhu, Lida, Pengsheng Xue, Qing Lan, Guiru Meng, Yuan Ren, Zhichao Yang, Peihua Xu, and Zhe Liu. "Recent research and development status of laser cladding: A review." *Optics & Laser Technology* 138 (2021): 106915.
15. Ahn, Dong-Gyu. "Directed energy deposition (DED) process: State of the art." *International Journal of Precision Engineering and Manufacturing-Green Technology* 8, no. 2 (2021): 703–742.
16. Yap, Chor Yen, Chee Kai Chua, Zhi Li Dong, Zhong Hong Liu, Dan Qing Zhang, Loong Ee Loh, and Swee Leong Sing. "Review of selective laser melting: Materials and applications." *Applied Physics Reviews* 2, no. 4 (2015):041101.
17. Squires, Lile, Ethan Roberts, and Amit Bandyopadhyay. "Radial bimetallic structures via wire arc directed energy deposition-based additive manufacturing." *Nature Communications* 14, no. 1 (2023): 3544.
18. Kumar, Nilesh, Het Bhavsar, P. V. S. Mahesh, Ashish Kumar Srivastava, Bhaskor J. Bora, Ambuj Saxena, and Amit Rai Dixit. "Wire arc additive manufacturing–a revolutionary method in additive manufacturing." *Materials Chemistry and Physics* 285 (2022): 126144.
19. Liu, Zhiying, Bei He, Tianyi Lyu, and Yu Zou. "A review on additive manufacturing of titanium alloys for aerospace applications: Directed energy deposition and beyond Ti-6Al-4V." *Jom* 73 (2021): 1804–1818.
20. Sarvankar, S. Ganesh, and Sanket Nandaram Yewale. "Additive manufacturing in automobile industry." *International Journal of Research in Aeronautical and Mechanical Engineering* 7, no. 4 (2019): 1–10.
21. Strickland, Jason D. "Applications of additive manufacturing in the marine industry." *Proceedings of PRADS2016* 4 (2016): 8.
22. Gudur, Srinath, Vishwanath Nagallapati, Sagar Pawar, Gopinath Muvvala, and Suryakumar Simhambhatla. "A study on the effect of substrate heating and cooling on bead geometry in wire arc additive manufacturing and its correlation with cooling rate." *Materials Today: Proceedings* 41 (2021): 431–436.
23. Singh, D. Dev, Thotakuri Mahender, and Avala Raji Reddy. "Powder bed fusion process: A brief review." *Materials Today: Proceedings* 46 (2021): 350–355.
24. Sutton, Austin T., Caitlin S. Kriewall, Ming C. Leu, and Joseph W. Newkirk. "Powder characterisation techniques and effects of powder characteristics on part properties in powder-bed fusion processes." *Virtual and Physical Prototyping* 12, no. 1 (2017): 3–29.
25. Mehrpouya, Mehrshad, Daniel Tuma, Tom Vaneker, Mohamadreza Afrasiabi, Markus Bambach, and Ian Gibson. "Multimaterial powder bed fusion techniques." *Rapid Prototyping Journal* 28, no. 11 (2022): 1–19.
26. McDonnell, Brian, Vito Errico, Paolo Posa, Andrea Angelastro, Agnieszka Furman, Eimear O'Hara, Sabina Luisa Campanelli, and Noel Harrison. "Bi-metallic lattice structures manufactured via an intralayer multi-material powder bed fusion method." *Additive Manufacturing* 89 (2024): 104301.
27. Rech, Joel, and Wit Grzesik. "New trends in hybrid finishing processes of metallic additively fabricated parts–a short review." *Journal of Machine Engineering* 23, no. 1 (2023): 36–46.
28. Grzesik, Wit. "Hybrid additive and subtractive manufacturing processes and systems: A review." *Journal of Machine Engineering* 18, no. 4 (2018): 5–24.

29. Grzesik, Wit, Adam Ruszaj, Wit Grzesik, and Adam Ruszaj. "Hybrid additive and subtractive processes." *Hybrid Manufacturing Processes: Physical Fundamentals, Modelling and Rational Applications* 118 (2021): 191–208.
30. Zhu, Zicheng, Vimal G. Dhokia, Aydin Nassehi, and Stephen T. Newman. "A review of hybrid manufacturing processes–state of the art and future perspectives." *International Journal of Computer Integrated Manufacturing* 26, no. 7 (2013): 596–615.
31. Smith, Scott, Tony Schmitz, Thomas Feldhausen, and Michael Sealy. "Hybrid metal additive/subtractive machine tools and applications." *CIRP Annals* 73, no. 2 (2024): 615–638.
32. Vellukunnel, Mathews T., Mukunda Khanal, and Xuejun Fan. "Thermal stresses in a bi layer assembly in electronics packaging." In *2023 24th International Conference on Thermal, Mechanical and Multi-Physics Simulation and Experiments in Microelectronics and Microsystems (EuroSimE)*, pp. 1–7. IEEE, 2023.
33. Zhu, Ziang, Renhai Shi, Andrew D. Klarner, Alan A. Luo, and Yiqing Chen. "Predicting and controlling interfacial microstructure of magnesium/aluminum bimetallic structures for improved interfacial bonding." *Journal of Magnesium and Alloys* 8, no. 3 (2020): 578–586.
34. Zixuan, L. I., Shahed Rezaei, W. A. N. G. Tao, H. A. N. Jianchao, S. H. U. Xuedao, Zbigniew Pater, and Qingxue Huang. "Recent advances and trends in roll bonding process and bonding model: A review." *Chinese Journal of Aeronautics* 36, no. 4 (2023): 36–74.
35. Liao-McPherson, Dominic, Efe C. Balta, Ryan Wüest, Alisa Rupenyan, and John Lygeros. "In-layer thermal control of a multi-layer selective laser melting process." In *2022 European Control Conference (ECC)*, pp. 1678–1683. IEEE, 2022.
36. Cai, Peng-Cheng, Guo-Hua Zhang, and Kuo-Chih Chou. "Simultaneously enhanced strength and ductility of W-Cu bimetallic composites assisted by continuous cellular dislocation structure." *International Journal of Plasticity* 184 (2025): 104188.
37. Gao, Lei, ChunGuo Xu, XueYuan Ge, QiPeng Hu, XiangLong Wang, Yuan Meng, ShiTeng Lu, and Bo Liu. "Study on microstructure and properties between alloy steel and powder steel via hot isostatic pressing diffusion bonding." *Materials Today Communications* 41 (2024): 110390.
38. Grzesik, Wit. *Advanced Machining Processes of Metallic Materials: Theory, Modelling and Applications*. Elsevier, 2008.
39. Bandyopadhyay, Amit, Yanning Zhang, and Bonny Onuike. "Additive manufacturing of bimetallic structures." *Virtual and Physical Prototyping* 17, no. 2 (2022): 256–294.
40. Peng, Xing, Lingbao Kong, Jerry Ying Hsi Fuh, and Hao Wang. "A review of post-processing technologies in additive manufacturing." *Journal of Manufacturing and Materials Processing* 5, no. 2 (2021): 38.
41. Simonetti, Sandra Kvaternik, Goran Turkalj, Damjan Banić, and Domagoj Lanc. "Bimetallic thin-walled box beam thermal buckling response." *Materials* 15, no. 21 (2022): 7537.
42. Bai, Shuang, and Jian Liu. "Additive manufacturing of bimetallic structures." *SN Applied Sciences* 2, no. 7 (2020): 1152.
43. Peng, Jiahao, Shijie Liu, Dong Wang, Anpeng Xu, Xin Huang, Tianqi Ma, Jing Wang, and Hang Li. "Design and optimization of thin-walled main support structure for space camera based on additive manufacturing." *Micromachines* 15, no. 2 (2024): 211.
44. Francis, Roger, Alan Turnbull, and Gareth Hinds. "Bimetallic Corrosion (Guides for Good Practice in Corrosion Control No. 5)." (2020).
45. Hirsch, Jürgen. "Recent development in aluminium for automotive applications." *Transactions of Nonferrous Metals Society of China* 24, no. 7 (2014): 1995–2002.
46. Hagarová, Mária, Janette Brezinová, Gabriela Baranová, Ján Viňáš, and Pavlo Maruschak. "Degradation of components in cars due to bimetallic corrosion." *Materials* 14, no. 12 (2021): 3323.

47. Parvizi, Mohsen, Aron Aladjem, and James Castle. "Behaviour of 90–10 cupronickel in sea water." *International Materials Reviews* 33, no. 1 (1988): 169–200.
48. Patin, Stanislav Aleksandrovich. *Environmental Impact of the Offshore Oil and Gas Industry*. Vol. 425. EcoMonitor Pub., 1999.
49. Shamsuddoha, Md, Md Mainul Islam, Thiru Aravinthan, Allan Manalo, and Kin-tak Lau. "Effectiveness of using fibre-reinforced polymer composites for underwater steel pipeline repairs." *Composite Structures* 100 (2013): 40–54.
50. Moosabeiki, Vahid, Afaq Khan, Mauricio Cruz Saldivar, Wim Van Paepegem, Brend P. Jonker, Eppo B. Wolvius, Jie Zhou, Nazli Tumer, Mohammad J. Mirzaali, and Amir A. Zadpoor. "Multi-material 3D printing of functionally graded soft-hard interfaces for enhancing mandibular kinematics of temporomandibular joint replacement prostheses." *Communications Materials* 5, no. 1 (2024): 1–11.
51. Appuhamillage, Gayan A., Sankalya S. Ambagaspitiya, Rohan S. Dassanayake, and Achintha Wijenayake. "3D and 4D printing of biomedical materials: Current trends, challenges, and future outlook." *Exploration of Medicine* 5, no. 1 (2024): 17–47.
52. Dada, Modupeola, and Patricia Popoola. "Recent advances in joining technologies of aluminum alloys: A review." *Discover Materials* 4, no. 1 (2024): 1–44.
53. Hassan, Hogr Z., and Najmadeen M. Saeed. "Advancements and applications of lightweight structures: A comprehensive review." *Discover Civil Engineering* 1, no. 1 (2024): 47.
54. Careri, Francesco, Raja H. U. Khan, Catherine Todd, and Moataz M. Attallah. "Additive manufacturing of heat exchangers in aerospace applications: A review." *Applied Thermal Engineering* 235 (2023): 121387.
55. Anirudhan, B. T., Jithin Devasia, Tejaswin Krishna, and Mebin T. Kuruvila. "Manufacturing of a bimetallic structure of stainless steel and mild steel through wire arc additive manufacturing—a critical review." *International Journal of Innovative Science and Research Technology* 5 (2020): 679–685.
56. Marchitto, Annalisa, Francesco Devia, Marco Fossa, Giovanni Guglielmini, and Corrado Schenone. "Experiments on two-phase flow distribution inside parallel channels of compact heat exchangers." *International Journal of Multiphase Flow* 34, no. 2 (2008): 128–144.
57. Liles, Kaitlin, and Ruth Amundsen. "NASA Passive Thermal Control Engineering Guidebook." (2023).
58. Yong, A. Y. L., Kar Heng Winston Seah, and Mustafizur Rahman. "Performance evaluation of cryogenically treated tungsten carbide tools in turning." *International Journal of Machine Tools and Manufacture* 46, no. 15 (2006): 2051–2056.
59. Li, Zhou, Jingwei Zhao, Hui Wu, Fanghui Jia, Ye Yao, Qingfeng Zhang, Sihai Jiao, and Zhengyi Jiang. "Experimental investigation on the mechanical and tribological coupled behaviour of bimetal composite under different states." *Surface Topography: Metrology and Properties* 7, no. 2 (2019): 025015.
60. Jien-Wei, Yeh. "Recent progress in high entropy alloys." *Annales de Chimie – Science des Matériaux* 31, no. 6 (2006): 633–648.
61. Zhang, Yong, Ting Ting Zuo, Zhi Tang, Michael C. Gao, Karin A. Dahmen, Peter K. Liaw, and Zhao Ping Lu. "Microstructures and properties of high-entropy alloys." *Progress in Materials Science* 61 (2014): 1–93.
62. Ron, Tomer, Amnon Shirizly, and Eli Aghion. "Additive manufacturing technologies of high entropy alloys (HEA): Review and prospects." *Materials* 16, no. 6 (2023): 2454.
63. Zhou, Yinghao, Zehuan Zhang, Dawei Wang, Weicheng Xiao, Jiang Ju, Shaofei Liu, Bo Xiao, Ming Yan, and Tao Yang. "New trends in additive manufacturing of high-entropy alloys and alloy design by machine learning: From single-phase to multiphase systems." *Journal of Materials Informatics* 2 (2022): 18.

Index

Note: **Bold** page numbers refer to tables and *italic* page numbers refer to figures.

R

S

T

For Product Safety Concerns and Information please contact our EU representative GPSR@taylorandfrancis.com Taylor & Francis Verlag GmbH, Kaufingerstraße 24, 80331 München, Germany